International Dictionary of Refrigeration

国际制冷词典

Dictionnaire International du Froid

International Institute of Refrigeration (IIR)
Institut International du Froid (IIF)

International Dictionary of Refrigeration

国际制冷词典

Dictionnaire International du Froid

CHINESE-ENGLISH-FRENCH: Terms

汉英法语术语

CHINOIS-ANGLAIS-FRANÇAIS：termes

Editor / Editeur scientifique: IIF-IIR

PEETERS

2008

International Institute of Refrigeration
Institut International du Froid
IIF-IIR, 177 boulevard Malesherbes, 75017 Paris, France
E-Mail: iif-iir@iifiir.org

PEETERS

ISBN: 978-90-429-2102-3 (Peeters Leuven)
ISBN: 978-2-7584-0037-0 (Peeters France)

D. 2008/0602/92

目录

前言

国际制冷词典共有 10 册组成，其中以英法册为基础册，包括英、法语的术语及其定义。另外 9 册分别为其它 9 种语言的术语和与之对应的英、法语术语组成。9 种语言分别为阿拉伯语、**汉语**（本册）、荷兰语、德语、意大利语、日语、挪威语、俄语和西班牙语。

一旦完成所有分册的出版，有望出版一本包含上述 11 种语言的国际制冷词典和交互式光盘。

本词典收录了大约 4400 个术语，其中 20% 为同义词。英法册收录了 3500 个术语的定义。而国际制冷学会（IIR）在 1975 年出版的国际制冷词典中仅收录 3000 个术语和 2400 个术语的定义。

本词典反映了 IIR 会员国语言的多样性。但 IIR 是以英语和法语作为官方语言。与 1975 年出版的国际制冷词典相对照，本词典增加了阿拉伯语，汉语和日语。

本词典还反映了制冷技术应用的多样性和发展。本词典包含了空气调节、低温生物学、制冷剂和液化气体等领域新出现的许多术语。同时还包括二章全新的内容：热泵、制冷与环境。更新了大部分现有术语的定义，并删除了 200 多条过时的术语。

整个词典的修订工作自 2000 年开始以来，历时 8 年。主要经历了以下几个阶段：

- 2000 年，选择关键词，制定词典的概要，设计词典的章节与 IIR 专业委员会的关联表，并将 1975 年版的词典录入电脑作为词典的基础；
  在准备工作期间，我们还参照了有关国际组织，尤其是国际标准化组织及一些欧、美团体和联合国某些机构对术语的定义。
- 2001~2002 年，以英语为工作语言，根据 IIR 的 10 个专业委员会以自愿为原则成立了 10 个工作组，并对术语进行更新；
- 2002~2003 年，10 个工作组成员对术语进行了英语定义；
- 2003~2005 年，包括汉语翻译组在内的 10 个翻译组对术语进行了 10 种语言的翻译。法语翻译组还翻译了术语的法语定义。总的来说，各有关国家制冷学会参与了整个翻译工作，但有 3 个例外：
  - 阿拉伯语是由 IIR 执行委员会的黎巴嫩代表和 IIR 的 7 个阿拉伯语国家代表共同合作完成的；
  - 西班牙语是由马德里制冷学会完成的；
  - 俄语是由 IIR 俄罗斯全国委员会完成的。
- 2005~2006 年，各专业委员会主席完成了审定工作；
- 2006~2007 年，出版前，工作组进行了二次审定，增加并翻译了 300 条近几年新出现的术语。

来自五大洲 30 多个国家的 IIR 的会员约 200 名专家，参与了本词典的修订工作。在此，我对他们付出的辛勤劳动致以最热烈最诚挚的感谢。同时对参与词典修订工作的 IIR 总部工作人员表示衷心的感谢，尤其是本词典修订项目的负责人副总干事 Daniel Viard，以及 Dany Furteau, Sylviane Fournier, Agnès Josserand-Broise 和 Susan Phalippou Mitchell。

本词典主要适用于众多的研究院所和用户，尤其是高校（机械、电力、化工和能源系等）以及制冷和空调领域的研究人员、工程师、技师、企业的情报部门和咨询行业人员。

同时，感谢 Peeters 出版商对本词典多个版本出版的大力支持。

最后，考虑到词典的定期更新、制冷技术的发展，将来可能会增加新的语言版本。

国际制冷词典的目标是成为从事制冷行业人员不可缺少的工具书。我们期待读者能够提供宝贵的意见和建议，以便使本词典成为更有价值的工具书。

Didier Coulomb, 国际制冷学会总干事

致谢

«国际制冷词典——**汉**英法语术语» 修订工作的专家名单

参与英语部分工作的 10 个**工作组**人员名单如下（10 个工作组根据 IIR 的 10 个专业委员会划分，由各专业委员会主席审定）：

A1 专业委员会：低温物理与低温工程

Ivan A. Arhharov	俄罗斯
Guobang Chen	中国
Guy Gistau Baguer – 协调人	法国
Philippe Lebrun – 审定人	法国
Rachid Rebiai	阿尔及利亚

A2 专业委员会：气体液化与分离

Rakesh Agraval	美国
Walter Castle – 协调人	英国
Václav Chrz – 审定人	捷克
Cyril Collins	英国
Wolfgang Foerg	德国
Krish R. Krishnamurthy	美国
Sebastian Muschelknautz	德国
Rachid Rebiai	阿尔及利亚
Vadim Udut	俄罗斯

B1 专业委员会:热力学及传递过程

Pradeep K. Bansal	新西兰
Jocelyn Bonjour	法国
Stefan Ditchev	保加利亚
Piotr Domanski – 协调人	美国
Carlos Alberto Infante Ferreira	荷兰
Dieter Gorenflo – 审定人	德国
Josef Ota	捷克
Milan Šafr	捷克
Oleg B. Tsvetkov	俄罗斯
Koichi Watanabe	日本
Felix Ziegler	德国

B2 专业委员会：制冷装置

Ben Adamson	澳大利亚
Clark Bullard – 审定人	美国
Carmine Casale	意大利

Mario Costantino	意大利
Daniele Farina	意大利
Ezio Fornasieri	意大利
Michael Kauffeld	德国
Grégoire Lambrinos	希腊
Siegfried Nowotny – 协调人	德国
Natividad López-Rodriguez	西班牙
Filippo de Rossi	意大利
Milan Šafr	捷克
Wolfgang Sussek	德国
Josef Weiss	奥地利
Felix Ziegler	德国

C1 专业委员会：低温生物学，低温医学

Anne Baudot	法国
Jean-Paul Homasson – 审定人	法国
Norio Murase	日本
Andreas Sputtek	德国

C2 专业委员会：食品科学及工程

Bryan Becker – 联合协调人	美国
Leif Bøgh Sørensen	丹麦
Christopher John Kennedy – 联合协调人	英国
Grégoire Lambrinos	希腊
Alain Le Bail	法国
Bart Nicolaï – 审定人	比利时
José Luis de la Plaza	西班牙
Rikuo Takai	日本
Josef Weiss	奥地利

D1 专业委员会： 冷藏

Milan Houska – 协调人	捷克
Manuel Lamúa Soldevilla	西班牙
Anders Lindborg	瑞典
Eleni Manolopoulou	希腊
Marek Michniewicz	波兰
David J. Tanner – 审定人	新西兰

D2 专业委员会：冷藏运输

Nevin Amos – 联合协调人	新西兰
Robert Heap – 联合协调人	英国
Sung Lim Kwon	美国
Hisashi Mitsufuji	日本
Girolamo Panozzo – 审定人	意大利
Rikuo Takai	日本
David J. Tanner	新西兰
Ruhe Xie	中国

E1 专业委员会：空气调节

Karel Hemzal	捷克
Peter Novak	斯洛文尼亚
Weiding Long – 审定人	中国
R.E. „Sam“ Luxton	澳大利亚
Pietro Mazzei – 协调人 (术语部分)	意大利
Jaroslav Wurm† – 协调人 (定义部分)	美国

E2 专业委员会：热泵，能量回收

Jos Bouma – 协调人	荷兰
Alberto Coronas Salcedo	西班牙
Daniele Farina	意大利
Hermann Halozan – 审定人	奥地利
Václav Havelský	斯洛伐克
Ahmed Kamoun	突尼斯
Petter Nekså	挪威
S. Srinivasa Murthy – 副协调人	印度

同时成立了 10 个语言翻译工作组。

在中国制冷学会的协调下，下列人员参与了汉语术语的翻译工作：

郭予信
洪朝生
郎四维
梁惊涛
刘作斌
马国远
邱忠岳（协调人）
徐庆磊
彦启森
俞炳丰（审定人）

在此，我们向参与汉语翻译工作的中国同僚和中国制冷学会所付出的辛勤劳动表示最诚挚的感谢！

在法国制冷学会的协调下，下列人员参与了法语术语及定义的翻译工作：

Mohamed Salah Abid (突尼斯)
Anne Baudot
Jean-Pierre Besancenot
Jocelyn Bonjour
François Clément – 协调人
Félix Depledt
Claude Dessaux
Maxime Duminil – 副协调人
Claude Ernst
Guy Gistau Baguer
Alain Goy
Jean-Paul Homasson
Alain Le Bail
Philippe Lebrun
Pierre Leclère – 副协调人
Bernard Lelièvre
Jean Rémy
Gérard Renaudin
Bernard Saulnier
Maurice Serrand
Georges Vrinat

使用说明

为了方便读者使用《国际制冷词典——汉英法语术语》，特作如下说明：

1. 本册的重点——术语部分，是根据专业分类并以字母序排列。

汉英法册共收录大约 4400 个术语。术语是由词或词组构成，其中 20% 的术语为同义词。

所有术语按专业类别分为 11 章。根据需要，每章又分为几节（第 8 章和第 11 章不分节）。所有术语均按字母序排列在 110 个小节中。

1.1. 组织结构与层次分类

在最底层的小节标题下，所有的术语按照英语术语的字母序排列。汉语术语放置在与其对应的英语术语和法语术语之间。

所有英语术语按照以下方式以标准印刷字体排列。
每个术语分配一个编号。如果术语出现在词典的奇数页，则该术语的编号放置在本页该术语所对应位置的右侧；反之，术语出现在词典的偶数页，该术语编号放置在本页该术语所对应位置的左侧。编号由术语所在的章节号和所在章节号的字母序号组成，中间用破折号将章节号和术语的字母序号分开，并且破折号后面的数字以**粗体字**印刷。例如：1.1.4-**3**.

J对于每个术语，其任一同义词都用*斜体字*印刷。
在每组同义词术语中，选出一个作为基本术语。
当基本术语作为同义词出现在一组术语中，基本术语以*斜体字*印刷，并且放在其它同义词的前面。

I在英法册中，英语术语的定义适用于英语基本术语及其同义词。与英语术语定义不同，法语术语的定义是由法语基本术语及其同义词的英语定义翻译而来的。

在本册中，与每组英语术语同义词对应的汉语基本术语用灰色底纹突出显示，其同义词不用灰色底纹显示。而与汉语术语对应的法语基本术语用标准字体印刷，其同义词用*斜体字*印刷。

大约有 40 个非常重要的英语术语（包括法语术语）放在 2 个不同的章节中。在英法册的英语索引中，该术语的编号是根据能够找到其定义的英语基本术语所在的章节来决定的。

1.2. 排版说明

– 在英语和法语中，术语分为 3 类：
  - 推荐使用的术语
  - 容许（或默许）术语（通俗用法）
  - 过时术语（可能继续使用）

后两类术语在整个英语术语中所占的比例不到 15%（分别为 11% 和 3%）。考虑到其所占的比例，在整个词典中包含了这两类术语：

◐ 容许（或默许）术语
○ 过时术语

当一个术语的同义词分属不同的类别时，将推荐使用的术语放在最前面，允许使用的术语次之，过时的术语放在最后面。

在汉语中，基本术语用灰色底纹突出显示，同义词不用灰色底纹显示，而上述提及的分类方式没有用在本册汉语术语中。

在所有语言分册中都用下面的例子说明上述的排版方式。在汉语册中，本例只有一个术语。如果有几个术语（同义词），它们会按照英语同义词和法语同义词的顺序排列。并且汉语的基本术语用灰色底纹突出显示，同义词不用灰色底纹显示。

boiler ○	发生器	générateur	2.3.3.1-**12**
generator		*bouilleur* ○	
regenerator			
desorber ◐			
desorber ◐	发生器	générateur	2.3.3.1-**13**
generator		*bouilleur* ○	
regenerator			
boiler ○			
generator	发生器	générateur	2.3.3.1-**19**
regenerator		*bouilleur* ○	
desorber ◐			
boiler ○			
regenerator	发生器	générateur	2.3.3.1-**27**
generator		*bouilleur* ○	
desorber ◐			
boiler ○			

- 根据使用需要，在英语字母序的索引中可找到英语基本术语的编号。
- 对于某些术语，会有下列的标示 (adj.) 或 (subst.)

  adj. = adjective (英语“形容词”) or adjectif (法语“形容词”)
  subst. = substantive (英语“名词”) or substantif (法语“名词”)
- 如果某个术语在某种语言中没有对应的术语翻译，将在对应的位置上放置 3 个星号“***”。

2. 英语术语的字母序索引

基于汉语的复杂性，提供汉语的索引难度非常大，本册使用英语术语的字母序索引提供如下功能：

- 在索引中，通过英语术语查找对应的汉语术语；
- 查阅词典中的汉语术语时，能够查到与之对应的英语术语。通过英语术语的索引，能够在英法册中查到英语的基本术语及其定义。

关于术语的英、法语定义，请查阅《国际制冷词典》基础册—英法册。

NOTES FOR READERS

In order to facilitate the use of the ***Chinese**-English-French* volume of the *International Dictionary of Refrigeration*, the following rationale was used in the presentation:

1. The core of the volume – containing the terms only – is organized primarily according to themes and is in alphabetical order.

The ***Chinese**-English-French* volume comprises approximately 4400 terms that are words or expressions composed of several words in the three languages; 20% of the terms are synonyms.

The terms are allocated to 11 chapters that are each subdivided into several sections of various levels of importance (only Chapters 8 and 11 are not subdivided). Within the 110 sections, the terms are classified in alphabetical order.

1.1. Organization and hierarchic classification

Under the smallest headings in the hierarchy, the terms are classified according to the alphabetical order of the English terms, and the Chinese terms are placed between the related English and French terms.

All the English terms in standard prints are arranged in this manner.
A code is assigned to each term and is placed on the right of odd-numbered pages and on the left of even-numbered pages. It is composed of the section number, separated by a dash from the rank of the term in the alphabetical order, within that particular section, and the latter number appears in **bold**: for example, 1.1.4-**3.**

For each term, any synonym appears in *italics.*
In each group of synonymous terms, one term has been selected as the main term.
When the main term appears as a synonym of one of the terms in the group, it appears in *italics,* again in first position, before the other synonyms.

In the *English-French* volume, the English definition referring to the main term and to its synonyms is mentioned with the main term and, opposite the main synonymous English term, the main French term and its synonyms are provided with the French definition relating to the English definition.

Opposite each term in a group of English synonyms, the main Chinese term with a grey background is given along with all its synonyms outside the grey zone, and the main French term is in standard print, along with its synonyms in *italics.*

About 40 English (and French) terms were deemed sufficiently important to be placed in two different sections. The code figuring in the English index is related to the main English term whose definition is to be found in one of these two sections of the *English-French* volume.

1.2. Presentation and typography

- We divided terms in English and French into three categories:
  - preferred terms;
  - tolerated terms (commonly used);
  - outdated terms (that may still be used).

The last two categories cover slightly less than 15% of the English terms (11% and 3% respectively) and we decided to mention the following after the terms involved:

◐ tolerated term
○ outdated term

When a term has synonyms in several categories, these categories are presented in the following order: preferred terms first, tolerated terms in second position and finally the outdated terms.

This distinction was not made in the case of Chinese terms; there is only a distinction between the main term with the grey background and the synonym(s) outside the grey zone.

Here is an example illustrating the principles described above (this example was selected for all the languages; here, in the case of the Chinese version, there is only one term; if there had been several terms, they would have been repeated in the same order opposite the English synonyms, as in the French version, with the main term in Chinese shown with the grey background and the synonyms outside the grey zone):

boiler ○	发生器	générateur	2.3.3.1-**12**
generator		*bouilleur* ○	
regenerator			
desorber ◐			
desorber ◐	发生器	générateur	2.3.3.1-**13**
generator		*bouilleur* ○	
regenerator			
boiler ○			
generator	发生器	générateur	2.3.3.1-**19**
regenerator		*bouilleur* ○	
desorber ◐			
boiler ○			
regenerator	发生器	générateur	2.3.3.1-**27**
generator		*bouilleur* ○	
desorber ◐			
boiler ○			

- If necessary, the code of the main English term can be found in the English alphabetical index.

- For certain terms, (adj.) or (subst.) are added:

  adj. = adjective (in English) or adjectif (in French)
  subst. = substantive (in English) or substantif (in French)

- If a term has no equivalent in a language, three asterisks *** are placed in the corresponding column.

2. Alphabetical Index of the English terms:

It proved to be impossible to generate an index of the Chinese characters, given the complex nature of the latter.

The alphabetical index of the English terms enables the following:

- by starting with the English term in the index, the corresponding Chinese term can be found;
- when the Chinese terms within the volume are consulted, the corresponding English terms are provided. By consulting the index of English terms, the main term in English can be found, and the definitions are provided in the *English-French* volume.

In order to access the definitions of the Chinese terms in English and French, it is essential to refer to the main *English-French* volume of the *International Dictionary of Refrigeration.*

AVIS AU LECTEUR

Afin de faciliter l'utilisation du volume des termes ***Chinois***-*anglais-français* du *Dictionnaire International du Froid*, les principes de présentation suivants ont été adoptés :

1. Le cœur du volume – avec les termes seulement – est organisé selon un classement thématique, puis alphabétique.

Le volume ***Chinois***-*anglais-français* est composé d'environ 4400 termes, qui sont des mots ou des expressions formées de plusieurs mots, dans les trois langues, 20 % des termes étant des synonymes.

Les termes de l'ouvrage sont regroupés dans 11 chapitres, qui peuvent être eux-mêmes subdivisés en plusieurs niveaux hiérarchiques de sous-chapitres (seuls les Chapitres 8 et 11 ne sont pas subdivisés). Au sein des 110 sous-chapitres, les termes sont classés par ordre alphabétique.

1.1. Organisation et hiérarchisation

A l'intérieur des plus petites subdivisions hiérarchiques, les termes sont classés selon l'ordre alphabétique des termes anglais, les termes chinois étant situés entre les termes anglais et français correspondants.

Tous les termes anglais, marqués en caractères standards, défilent dans cet ordre.
A chaque terme est associé un code qui figure dans la colonne située à droite sur les pages impaires et à gauche sur les pages paires. Il est constitué du numéro du sous-chapitre séparé par un trait d'union du numéro d'ordre alphabétique du terme dans le sous-chapitre, ce dernier numéro étant écrit en **gras** : par exemple, 1.1.4-**3**.

Chacun des termes est systématiquement accompagné de ses synonymes éventuels marqués en *italique*.
Dans chaque groupe de termes synonymes, l'un d'entre eux a été choisi comme terme principal.
Lorsque le terme principal est en position de synonyme par rapport à un autre terme du groupe, il est écrit en *italique*, toujours en tête des autres synonymes.

Dans le volume *anglais-français*, la définition anglaise correspondant au terme principal et à ses synonymes est mentionnée avec le terme principal et, en face du terme principal anglais homologue, le terme principal français et ses synonymes sont accompagnés de la définition française correspondant à la définition anglaise.

En face de chacun des termes d'un groupe de synonymes anglais, on retrouve, à l'intérieur d'une bande grisée, le terme principal chinois accompagné, à l'extérieur de la bande grisée, de l'ensemble de ses synonymes, et le terme principal français, en caractères standards, accompagné de ses synonymes en *italique*.

Une quarantaine de termes anglais (et français) ont été considérés comme suffisamment importants pour être présentés dans deux sous-chapitres différents. Le code retenu dans l'index anglais correspond au code du terme anglais principal dont on a mis la définition dans l'un des deux sous-chapitres du volume *anglais-français*.

1.2. Présentation et typographie

– Nous avons distingué trois catégories de termes en anglais et en français :
  - les termes préférables,
  - les termes tolérés (d'usage courant),
  - les termes désuets (susceptibles d'être encore utilisés).

Les deux dernières catégories représentant un peu moins de 15 % des termes anglais (respectivement 11 % et 3 %), nous avons choisi d'adjoindre la marque suivante après chacun des termes concernés :

◐ terme toléré
○ terme désuet

Lorsqu'un terme est associé à des synonymes appartenant à plusieurs catégories, ceux-ci sont présentés dans l'ordre suivant : les termes préférables en premier, les termes tolérés en deuxième et les termes désuets en dernier.

Cette distinction n'a pas été faite avec les termes chinois ; il y a seulement la distinction entre le terme principal à l'intérieur de la bande grisée et le (ou les) synonyme(s) à l'extérieur de la bande grisée.

Exemple illustrant les principes énoncés précédemment (cet exemple a été retenu pour toutes les langues ; ici, dans le cas du chinois, il n'y a qu'un seul terme ; s'il y en avait plusieurs, ils seraient répétés dans le même ordre en face des synonymes anglais, comme pour le français, à la seule différence que le terme principal en chinois serait à l'intérieur de la bande grisée et que ses synonymes seraient à l'extérieur de la bande grisée) :

boiler ○	发生器	générateur	2.3.3.1-**12**
generator		*bouilleur* ○	
regenerator			
desorber ◐			
desorber ◐	发生器	générateur	2.3.3.1-**13**
generator		*bouilleur* ○	
regenerator			
boiler ○			
generator	发生器	générateur	2.3.3.1-**19**
regenerator		*bouilleur* ○	
desorber ◐			
boiler ○			
regenerator	发生器	générateur	2.3.3.1-**27**
generator		*bouilleur* ○	
desorber ◐			
boiler ○			

– Si nécessaire, on trouvera dans l'index alphabétique anglais le code du terme principal anglais ;

– Certains termes anglais et français peuvent être accompagnés de la mention : (adj.) ou (subst.) ;

  adj. = adjective (en anglais) ou adjectif (en français)
  subst. = substantive (en anglais) ou substantif (en français) ;

– Si un terme n'a pas d'équivalent dans une langue, trois astérisques *** sont placées dans la colonne correspondante.

2. Index alphabétique des termes anglais :

Il n'a pas été possible de faire un index des termes chinois, compte-tenu de la complexité de ce travail.

L'index alphabétique des termes anglais permet de faire au moins deux choses :

- On peut partir d'un terme anglais dans l'index pour trouver le terme chinois correspondant,
- On peut partir d'un terme chinois dans le cœur du volume, auquel cas on a le ou les termes anglais correspondants ; et, si nécessaire, la consultation de l'index des termes anglais permet d'avoir le terme principal anglais, dont on a la définition dans le volume *anglais-français*.

Pour avoir accès aux définitions des termes chinois en anglais et en français, il est indispensable de se référer au volume *anglais-français*, volume de base du *Dictionnaire International du Froid*.

版本说明

根据语言安排的需要，国际制冷词典由以下十个分册组成：

- 英法册（基础册），包括英、法语的术语及其定义，以及以英文字母序和法文字母序排列的索引。
- 9 种不同语言分册，包含其一种语言的术语和与之对应的英、法语术语：
  - 法语和英语索引的阿拉伯语分册，
  - 英语索引的 **汉语分册**，
  - 荷兰语索引的荷兰语分册，
  - I德语索引的德语分册，
  - 意大利语索引的意大利分册，
  - 英语索引的日语分册，
  - 挪威语索引的挪威语分册，
  - 俄语索引的俄语分册，
  - 西班牙语索引的西班牙语分册。

一旦完成所有分册的出版，有望出版一本包含上述 11 种语言的国际制冷词典和交互式光盘。

章 1. 基本知识

◐ 容许（或默许）术语

○ 过时术语

ENGLISH	汉语	FRANÇAIS	
SECTION 1.1 *General concepts and instrumentation* **SECTION 1.1.1** *General concepts and instrumentation: general background*	节 **1.1** 常用概念与测量 节 **1.1.1** 常用概念与测量：基础知识	**SOUS-CHAPITRE 1.1** *Concepts généraux et instrumentation* **SOUS-CHAPITRE 1.1.1** *Généralités sur les concepts généraux et l'instrumentation*	
authorized person	专职人员 指定 审定人员	personne autorisée	**1.1.1-1**
British thermal unit (Btu) ◐	英热单位（Btu）	Btu ◐	**1.1.1-2**
cold (adj.)	冷的	froid (adj.)	**1.1.1-3**
cold chain	冷链	chaîne du froid	**1.1.1-4**
competence	胜任	compétence	**1.1.1-5**
congeal (to) *freeze (to)*	冻结（动词）	geler	**1.1.1-6**
cooling	冷却	refroidissement	**1.1.1-7**
critical density	临界密度	masse volumique critique	**1.1.1-8**
density	密度 容重	masse volumique	**1.1.1-9**
food chain	食品链	chaîne alimentaire	**1.1.1-10**
freeze (to) *congeal (to)*	冻结（动词）	geler	**1.1.1-11**
frost	霜	givre	**1.1.1-12**
(graduate) refrigerating engineer ◐ *(graduate) refrigeration engineer*	制冷工程师（取得学位的）	ingénieur frigoriste	**1.1.1-13**
(graduate) refrigeration engineer *(graduate) refrigerating engineer ◐*	制冷工程师（取得学位的）	ingénieur frigoriste	**1.1.1-14**
heat capacity *thermal capacity*	热容 热容量	capacité thermique	**1.1.1-15**
joule (J)	焦耳（J）	joule (J)	**1.1.1-16**
kilocalorie (kcal) ◐	千卡（kcal）	kilocalorie (kcal) ◐ *millithermie (mth)* ○	**1.1.1-17**
newton (N)	牛顿（N）	newton (N)	**1.1.1-18**
pascal (Pa)	帕（Pa）	pascal (Pa)	**1.1.1-19**
reference value	基准值	valeur de référence	**1.1.1-20**
refrigerate (to)	制冷（动词）	refroidir	**1.1.1-21**
refrigerating	制冷的	frigorifique (adj.)	**1.1.1-22**
refrigerating capacity	制冷功率（量） 产冷量	puissance frigorifique	**1.1.1-23**
refrigeration	制冷	froid (artificiel)	**1.1.1-24**
refrigeration contractor	制冷工程承包商	entrepreneur frigoriste	**1.1.1-25**
refrigeration engineer ◐ *refrigeration mechanic*	制冷技师	mécanicien frigoriste	**1.1.1-26**
refrigeration engineering	制冷工程	génie frigorifique *technique frigorifique* *technique du froid*	**1.1.1-27**

	ENGLISH	汉语	FRANÇAIS
1.1.1-**28**	refrigeration installation engineer	制冷安装技师	monteur frigoriste
1.1.1-**29**	refrigeration mechanic *refrigeration engineer* ◐	制冷技师	mécanicien frigoriste
1.1.1-**30**	refrigeration serviceman *refrigeration technician* ◐	制冷维修技工 制冷技术员	dépanneur frigoriste
1.1.1-**31**	refrigeration technician ◐ *refrigeration serviceman*	制冷维修技工 制冷技术员	dépanneur frigoriste
1.1.1-**32**	relative molecular mass	相对分子质量	masse moléculaire relative
1.1.1-**33**	specific gravity ◐	比重	densité ◐
1.1.1-**34**	specific heat c *specific heat capacity* c	比热容 c	capacité thermique massique c *chaleur massique* c ○ *chaleur spécifique* c ○
1.1.1-**35**	specific heat capacity c *specific heat* c	比热容 c	capacité thermique massique c *chaleur massique* c ○ *chaleur spécifique* c ○
1.1.1-**36**	specific heat capacity c_p at constant pressure	定压比热容 c_p	capacité thermique massique c_p à pression constante *chaleur massique à pression constante* c_p ○ *chaleur spécifique à pression constante* c_p ○
1.1.1-**37**	specific heat capacity c_v at constant volume	定容比热容 c_v	capacité thermique massique c_v à volume constant *chaleur massique à volume constant* c_v ○ *chaleur spécifique à volume constant* c_v ○
1.1.1-**38**	specific heat ratio c_p/c_v	比热容比 c_p/c_v	rapport des capacités thermiques massiques c_p/c_v
1.1.1-**39**	specific volume	比体积 比容	volume massique
1.1.1-**40**	therm ○	热量单位（色姆）	***
1.1.1-**41**	thermal capacity *heat capacity*	热容 热容量	capacité thermique
1.1.1-**42**	ton of refrigeration (T.R.) ◐	美国冷吨 （T. R. ）	***
1.1.1-**43**	watt (W)	瓦 （W）	watt (W)
1.1.1-**44**	***	(负)千卡	frigorie (fg) ○
1.1.1-**45**	***	法国热量单位 （th）	thermie (th) ○

	SECTION 1.1.2 *Temperature*	节 **1.1.2** 温度	**SOUS-CHAPITRE 1.1.2** *Température*
1.1.2-**1**	absolute temperature *thermodynamic temperature*	绝对温度 热力学温度	température absolue *température thermodynamique*
1.1.2-**2**	absolute zero	绝对零度	zéro absolu
1.1.2-**3**	acoustical thermometer	声学温度计	thermomètre acoustique
1.1.2-**4**	alcohol thermometer	酒精温度计	thermomètre à alcool
1.1.2-**5**	ambient temperature	环境温度	température ambiante
1.1.2-**6**	arithmetic mean temperature difference	算术平均温差	écart moyen arithmétique de température

ENGLISH	汉语	FRANÇAIS	
average temperature *mean temperature*	平均温度	température moyenne	1.1.2-**7**
bimetallic element *bimetallic strip*	双金属元件 双金属片	bilame	1.1.2-**8**
bimetallic strip *bimetallic element*	双金属元件 双金属片	bilame	1.1.2-**9**
bimetallic thermometer	双金属片温度计	thermomètre à bilame	1.1.2-**10**
Celsius (temperature) scale	摄氏温标	échelle (thermométrique) Celsius	1.1.2-**11**
Celsius temperature	摄氏温度	température Celsius	1.1.2-**12**
Curie point	居里温度	point de Curie	1.1.2-**13**
degree Celsius (°C)	摄氏温度 （℃）	degré Celsius (°C)	1.1.2-**14**
degree Fahrenheit (°F) ○	华氏温度 （℉）	degré Fahrenheit (°F) ○	1.1.2-**15**
degree Rankine (°R) ○	兰氏温度 （°R）	degré Rankine (°R) ○	1.1.2-**16**
dial thermometer	指针温度计	thermomètre à cadran	1.1.2-**17**
distant reading thermometer ○ *remote reading thermometer*	遥测温度计	téléthermomètre *thermomètre à distance* ○	1.1.2-**18**
electric contact thermometer	电触点温度计	thermomètre à contacts électriques	1.1.2-**19**
external temperature *outside temperature*	外界温度	température extérieure	1.1.2-**20**
Fahrenheit (temperature) scale ◐	华氏温标	échelle (thermométrique) Fahrenheit	1.1.2-**21**
gas thermometer	气体温度计	thermomètre à gaz	1.1.2-**22**
indicating thermometer	指示温度计	thermomètre à lecture directe	1.1.2-**23**
inside temperature *internal temperature*	内部温度	température intérieure	1.1.2-**24**
internal temperature *inside temperature*	内部温度	température intérieure	1.1.2-**25**
IPTS-68	国际实用温标-1968	IPTS-68	1.1.2-**26**
ITS-90	国际温标-1990	ITS-90	1.1.2-**27**
kelvin (K)	开尔文（K）	kelvin (K)	1.1.2-**28**
Kelvin (temperature) scale *thermodynamic temperature scale*	开氏温标 热力学温标	échelle (thermométrique) Kelvin *échelle de température thermodynamique*	1.1.2-**29**
limiting temperature	极限温度	température limite	1.1.2-**30**
logarithmic mean temperature difference	对数平均温差	écart moyen logarithmique de température	1.1.2-**31**
lowering of temperature *temperature reduction*	温度下降 温降	abaissement de température *réduction de température*	1.1.2-**32**
magnetic temperature	磁温度	température magnétique	1.1.2-**33**
magnetic thermometer	磁温度计	thermomètre magnétique	1.1.2-**34**
maximum service temperature	最高使用温度	température maximale de service	1.1.2-**35**
mean temperature *average temperature*	平均温度	température moyenne	1.1.2-**36**
mean temperature difference	平均温差	écart moyen de température	1.1.2-**37**
mercury thermometer	水银温度计	thermomètre à mercure	1.1.2-**38**
minimum service temperature	最低使用温度	température minimale de service	1.1.2-**39**

	ENGLISH	汉语	FRANÇAIS
1.1.2-**40**	Néel point *Néel temperature* ◐	奈耳温度	point de Néel
1.1.2-**41**	Néel temperature ◐ *Néel point*	奈耳温度	point de Néel
1.1.2-**42**	nuclear resonance thermometer	核磁共振温度计	thermomètre à résonance nucléaire
1.1.2-**43**	operating temperature *working temperature* ◐	运行温度	température de fonctionnement
1.1.2-**44**	outside temperature *external temperature*	外界温度	température extérieure
1.1.2-**45**	pyrometer	高温计	pyromètre
1.1.2-**46**	Rankine temperature scale ◐	兰氏温标	échelle thermométrique Rankine ◐
1.1.2-**47**	recording thermometer *temperature recorder* *thermograph* ◐	温度记录仪	thermographe *thermomètre enregistreur*
1.1.2-**48**	reference mean temperature	基准平均温度	température moyenne de référence
1.1.2-**49**	remote reading thermometer *distant reading thermometer* ○	遥测温度计	téléthermomètre *thermomètre à distance* ○
1.1.2-**50**	resistance thermometer	电阻温度计	thermomètre à résistance
1.1.2-**51**	solid expansion thermometer	固体膨胀温度计	thermomètre à dilatation solide
1.1.2-**52**	standard ambient temperature	标准环境温度	température normale ambiante
1.1.2-**53**	subcooled refrigerant temperature	过冷制冷剂温度	température d'un fluide frigorigène sous-refroidi
1.1.2-**54**	surface temperature	表面温度	température de surface
1.1.2-**55**	temperature	温度	température
1.1.2-**56**	temperature difference	温差	différence de température *écart de température*
1.1.2-**57**	temperature drop	温降	chute de température
1.1.2-**58**	temperature fluctuation	温度波动	fluctuation de température
1.1.2-**59**	temperature gradient	温度梯度	gradient de température
1.1.2-**60**	temperature increment	温度增量	accroissement de température
1.1.2-**61**	temperature profile	温度分布图	profil des températures
1.1.2-**62**	temperature recorder *recording thermometer* *thermograph* ◐	温度记录仪	thermographe *thermomètre enregistreur*
1.1.2-**63**	temperature reduction *lowering of temperature*	温度下降 温降	abaissement de température *réduction de température*
1.1.2-**64**	temperature rise	温升	élévation de température
1.1.2-**65**	temperature scale	温标	échelle de température *échelle thermométrique*
1.1.2-**66**	temperature variation	温度变化	variation de température
1.1.2-**67**	thermal convector ○ *thermocouple* *thermo-junction* ◐ *thermoelectric couple* ◐	热电偶	couple thermoélectrique *thermocouple*
1.1.2-**68**	thermistor	热敏电阻	thermistance

ENGLISH	汉语	FRANÇAIS	
thermo-junction ◐ *thermocouple* *thermoelectric couple* ◐ *thermal convector* ○	热电偶	couple thermoélectrique *thermocouple*	1.1.2-**69**
thermocouple *thermo-junction* ◐ *thermoelectric couple* ◐ *thermal convector* ○	热电偶	couple thermoélectrique *thermocouple*	1.1.2-**70**
thermodynamic temperature *absolute temperature*	绝对温度 热力学温度	température absolue *température thermodynamique*	1.1.2-**71**
thermodynamic temperature scale *Kelvin (temperature) scale*	开氏温标 热力学温标	échelle (thermométrique) Kelvin *échelle de température thermodynamique*	1.1.2-**72**
thermoelectric couple ◐ *thermocouple* *thermo-junction* ◐ *thermal convector* ○	热电偶	couple thermoélectrique *thermocouple*	1.1.2-**73**
thermograph ◐ *recording thermometer* *temperature recorder*	温度记录仪	thermographe *thermomètre enregistreur*	1.1.2-**74**
thermometer	温度计	thermomètre	1.1.2-**75**
thermometry	检温学	thermométrie	1.1.2-**76**
thermopile	热电堆 热（温差）电偶 热（温差）电池	thermopile	1.1.2-**77**
vapour pressure thermometer	蒸气压温度计	thermomètre à tension de vapeur	1.1.2-**78**
working temperature ◐ *operating temperature*	运行温度	température de fonctionnement	1.1.2-**79**

SECTION 1.1.3 *Pressure*	节 **1.1.3** 压力	**SOUS-CHAPITRE 1.1.3** *Pression*	
absolute pressure	绝对压力	pression absolue	1.1.3-**1**
absolute pressure gauge	绝对压力表	manomètre absolu	1.1.3-**2**
absolute vacuum	绝对真空	vide absolu	1.1.3-**3**
annular chamber	环状容器	chambre annulaire *espace annulaire*	1.1.3-**4**
atmospheric pressure	大气压	pression atmosphérique	1.1.3-**5**
barometer	气压计	baromètre	1.1.3-**6**
bellows	波纹管	soufflet	1.1.3-**7**
Bourdon gauge	波登管压力计	manomètre de Bourdon	1.1.3-**8**
combined pressure and vacuum gauge *compound gauge*	复合真空压力表	manovacuomètre	1.1.3-**9**
compound gauge *combined pressure and vacuum gauge*	复合真空压力表	manovacuomètre	1.1.3-**10**
control pressure gauge	压力控制表	manomètre à fonction(s) de commande	1.1.3-**11**
diaphragm	膜片	membrane	1.1.3-**12**
diaphragm manometer *membrane manometer*	膜片式压力计	manomètre à membrane	1.1.3-**13**

	ENGLISH	汉语	FRANÇAIS
1.1.3-**14**	differential pressure	压差	pression différentielle
1.1.3-**15**	differential pressure gauge	压差式压力计	manomètre différentiel
1.1.3-**16**	duplex pressure gauge	复式压力计	manomètre duplex
1.1.3-**17**	dynamic head *velocity head*	动压头	charge dynamique *hauteur dynamique*
1.1.3-**18**	dynamic pressure *velocity pressure*	动压	pression dynamique *pression due à la vitesse*
1.1.3-**19**	edgewise pressure gauge	边缘读数式压力计	manomètre de profil
1.1.3-**20**	excess pressure	超压	surpression
1.1.3-**21**	gauge pressure	表压	pression relative *pression effective*
1.1.3-**22**	hydrostatic pressure	液体静压力	pression hydrostatique
1.1.3-**23**	indicating pressure gauge	直读压力表	manomètre indicateur
1.1.3-**24**	indicating pressure gauge with an elastic measuring element	弹性管指示压力表	manomètre métallique indicateur
1.1.3-**25**	ionization vacuum gauge	电离真空表	jauge à ionisation *manomètre à ionisation*
1.1.3-**26**	liquid-filled pressure gauge	充液式压力表	manomètre à liquide amortisseur
1.1.3-**27**	liquid inlet pressure	液体进口压力	pression d'entrée du fluide
1.1.3-**28**	(liquid level) manometer *U-tube manometer*	U 型管压力计	manomètre à tube de liquide
1.1.3-**29**	liquid outlet pressure	液体出口压力	pression de sortie du fluide frigoporteur
1.1.3-**30**	loss of head	压头损失	perte de charge
1.1.3-**31**	membrane manometer *diaphragm manometer*	膜片式压力计	manomètre à membrane
1.1.3-**32**	osmotic pressure	渗透压	pression osmotique
1.1.3-**33**	partial pressure	分压力	pression partielle
1.1.3-**34**	piezometer ring	环形流压力计	bague piézométrique
1.1.3-**35**	pointer	指针	aiguille
1.1.3-**36**	pressure	压力	pression
1.1.3-**37**	pressure drop	压力降	chute de pression
1.1.3-**38**	pressure gauge	压力表	manomètre
1.1.3-**39**	pressure loss	压力损失	perte de pression
1.1.3-**40**	pressure-responsive element	压力敏感元件	organe moteur
1.1.3-**41**	refrigerant inlet pressure	制冷剂进口压力	pression d'entrée du fluide frigorigène
1.1.3-**42**	relative vapour pressure	相对蒸气压力	pression relative de vapeur
1.1.3-**43**	safety pattern gauge	安全型压力表	manomètre de sureté
1.1.3-**44**	service gauge	维修压力表	manomètre de monteur
1.1.3-**45**	shank	压力表体	raccord
1.1.3-**46**	stagnation pressure	滞止压力	pression d'arrêt
1.1.3-**47**	standard atmospheric pressure	标准大气压	pression atmosphérique normale
1.1.3-**48**	static head	静压头	hauteur statique *charge statique*
1.1.3-**49**	static pressure	静压	pression statique

ENGLISH	汉语	FRANÇAIS	
thermal conductivity vacuum gauge	导热真空计	jauge thermique *manomètre thermique*	1.1.3-**50**
total head	全压头	charge totale	1.1.3-**51**
total pressure	全压	pression totale	1.1.3-**52**
U-tube manometer *(liquid level) manometer*	U 型管压力计	manomètre à tube de liquide	1.1.3-**53**
ultra-high vacuum	超高真空	ultravide	1.1.3-**54**
vacuum	真空	vide *gaz raréfié*	1.1.3-**55**
vacuum gauge	真空计	jauge à vide *manomètre à vide*	1.1.3-**56**
vapour pressure	蒸气压	pression de vapeur	1.1.3-**57**
velocity head *dynamic head*	动压头	charge dynamique *hauteur dynamique*	1.1.3-**58**
velocity pressure *dynamic pressure*	动压	pression dynamique *pression due à la vitesse*	1.1.3-**59**
viscosity manometer	粘度压力计	manomètre à viscosité	1.1.3-**60**
wall (pressure) tapping	管壁（压）孔	prise (de pression) à la paroi	1.1.3-**61**
water column	水柱	colonne d'eau	1.1.3-**62**

SECTION 1.1.4 *Humidity*	节 **1.1.4** 湿度	**SOUS-CHAPITRE 1.1.4** *Humidité*	
absolute humidity *water vapour concentration* ○	绝对湿度 水蒸汽密度	humidité absolue *concentration de la vapeur d'eau* ○	1.1.4-**1**
absorption hygrometer *chemical hygrometer*	吸收式湿度计	hygromètre à absorption	1.1.4-**2**
air dry-bulb temperature *dry-bulb temperature*	空气干球温度 干球温度	température de bulbe sec *température sèche (de l'air)* ◐	1.1.4-**3**
air wet-bulb temperature *wet-bulb temperature*	空气湿球温度 湿球温度	température de bulbe humide *température humide (de l'air)* ◐	1.1.4-**4**
aspirated hygrometer *aspiration psychrometer*	通风式干湿球湿度计	psychromètre à aspiration *psychromètre ventilé*	1.1.4-**5**
aspiration psychrometer *aspirated hygrometer*	通风式干湿球湿度计	psychromètre à aspiration *psychromètre ventilé*	1.1.4-**6**
chemical hygrometer *absorption hygrometer*	吸收式湿度计	hygromètre à absorption	1.1.4-**7**
degree of saturation *saturation ratio*	饱和度	degré de saturation	1.1.4-**8**
dehumidification	减（除）湿	déshumidification	1.1.4-**9**
dehumidify (to)	减（除）湿(动词)	déshumidifier	1.1.4-**10**
dehydrate (to)	脱水(动词)	déshydrater	1.1.4-**11**
dehydration *drying*	脱水 干燥	déshydratation *séchage*	1.1.4-**12**
desiccation	干燥	dessiccation	1.1.4-**13**
dew (1)	结露	rosée	1.1.4-**14**

	ENGLISH	汉语	FRANÇAIS
1.1.4-**15**	dew (2)	露水	rosée
1.1.4-**16**	dew point *dew-point (temperature)*	露点(温度)	température de rosée
1.1.4-**17**	dew-point depression	露点温差	écart du point de rosée
1.1.4-**18**	dew-point hygrometer	露点湿度计	hygromètre à point de rosée
1.1.4-**19**	dew-point (temperature) *dew point*	露点(温度)	température de rosée
1.1.4-**20**	dry air	干空气	air sec
1.1.4-**21**	dry bulb	干球	bulbe sec
1.1.4-**22**	dry-bulb temperature *air dry-bulb temperature*	空气干球温度 干球温度	température de bulbe sec *température sèche (de l'air)* ◐
1.1.4-**23**	dry-bulb thermometer	干球温度计	thermomètre (à bulbe) sec
1.1.4-**24**	drying *dehydration*	脱水 干燥	séchage *déshydratation*
1.1.4-**25**	electrical hygrometer	电测湿度计	hygromètre électrique
1.1.4-**26**	electrolytic hygrometer	电解湿度计	hygromètre électrolytique
1.1.4-**27**	fog *mist*	雾	brouillard *buée*
1.1.4-**28**	frost point	结霜温度	point de gelée blanche *point de givre*
1.1.4-**29**	glaze	薄冰层	verglas *givre transparent*
1.1.4-**30**	hair hygrometer	毛发湿度计	hygromètre à cheveux
1.1.4-**31**	hoarfrost	白霜	gelée blanche
1.1.4-**32**	humid *moist*	潮湿	humide
1.1.4-**33**	humid air	湿空气	air humide
1.1.4-**34**	humid volume ○	湿体积	volume spécifique *volume massique rapporté à l'air sec*
1.1.4-**35**	humidification	加湿	humidification
1.1.4-**36**	humidify (to)	加湿(动词)	humidifier
1.1.4-**37**	humidity	湿度	humidité
1.1.4-**38**	humidity ratio *mixing ratio* *moisture content (1)*	含湿量	humidité spécifique (air humide)
1.1.4-**39**	hygrometer	湿度计	hygromètre
1.1.4-**40**	hygrometry	测湿学	hygrométrie
1.1.4-**41**	hygroscopic	吸湿的	hygroscopique
1.1.4-**42**	mist *fog*	雾	buée *brouillard*
1.1.4-**43**	mixing ratio *humidity ratio* *moisture content (1)*	含湿量	humidité spécifique (air humide)
1.1.4-**44**	moist *humid*	潮湿	humide

ENGLISH	汉语	FRANÇAIS	
moisture carry-over	载湿	entraînement de gouttelettes (d'eau)	1.1.4-**45**
moisture content (1) *humidity ratio* *mixing ratio*	含湿量	humidité spécifique (air humide)	1.1.4-**46**
moisture content (2) *water content*	含水量	teneur en eau (substance solide)	1.1.4-**47**
moisture indicator	湿度指示器	indicateur d'humidité	1.1.4-**48**
moisture transfer	传湿	transfert d'humidité *transport d'humidité*	1.1.4-**49**
mole fraction of the water vapour	水蒸汽的克分子比	fraction molaire de vapeur d'eau *titre molaire de vapeur d'eau* ◐	1.1.4-**50**
organic hygrometer	有机湿度计	hygromètre organique	1.1.4-**51**
psychrometer	干湿球温度计	psychromètre	1.1.4-**52**
psychrometric chart	温湿图	diagramme psychrométrique	1.1.4-**53**
psychrometrics *psychrometry*	测湿法	psychrométrie	1.1.4-**54**
psychrometry *psychrometrics*	测湿法	psychrométrie	1.1.4-**55**
relative humidity	相对湿度	humidité relative	1.1.4-**56**
relative humidity with respect to ice (W.M.O.)	相对于冰的相对湿度 （W. M. O. ）	humidité relative par rapport à la glace (O.M.M.)	1.1.4-**57**
relative humidity with respect to water (W.M.O.) ◐	相对于水的相对湿度 （W. M. O. ）	humidité relative par rapport à l'eau (O.M.M.) ◐	1.1.4-**58**
rime	凇（白）霜	givre blanc	1.1.4-**59**
saturated air ◐ *saturated humid air*	饱和空气	air humide saturé *air saturé* ◐	1.1.4-**60**
saturated humid air *saturated air* ◐	饱和空气	air humide saturé *air saturé* ◐	1.1.4-**61**
saturation ratio *degree of saturation*	饱和度	degré de saturation	1.1.4-**62**
sling hygrometer *sling psychrometer*	手摇干湿球温度计	psychromètre à rotation *psychromètre fronde*	1.1.4-**63**
sling psychrometer *sling hygrometer*	手摇干湿球温度计	psychromètre à rotation *psychromètre fronde*	1.1.4-**64**
specific humidity *water vapour content* ◐	含湿度 水蒸汽含量	humidité spécifique *teneur en vapeur d'eau* ◐	1.1.4-**65**
steam *water vapour*	水蒸汽	vapeur d'eau	1.1.4-**66**
supersaturated air	过饱和空气	air sursaturé	1.1.4-**67**
sweating	表面凝水	condensation d'eau (sur une surface)	1.1.4-**68**
water content *moisture content (2)*	含水量	teneur en eau (substance solide)	1.1.4-**69**
water vapour *steam*	水蒸汽	vapeur d'eau	1.1.4-**70**
water vapour concentration ○ *absolute humidity*	绝对湿度 水蒸汽密度	humidité absolue *concentration de la vapeur d'eau* ○	1.1.4-**71**
water vapour content ◐ *specific humidity*	含湿度 水蒸汽含量	humidité spécifique *teneur en vapeur d'eau* ◐	1.1.4-**72**

	ENGLISH	汉语	FRANÇAIS
1.1.4-**73**	wet bulb	湿球	bulbe humide
1.1.4-**74**	wet-bulb depression	干湿球温差	différence psychrométrique
1.1.4-**75**	wet-bulb temperature *air wet-bulb temperature*	空气湿球温度 湿球温度	température de bulbe humide *température humide (de l'air)* ◐
1.1.4-**76**	wet-bulb thermometer	湿球温度计	thermomètre (à bulbe) humide

	SECTION 1.1.5 *Heat*	节 **1.1.5** 热，热量	**SOUS-CHAPITRE 1.1.5** *Chaleur*
1.1.5-**1**	amount of heat *heat quantity*	热量	quantité de chaleur
1.1.5-**2**	area coefficient of heat loss	热损失面积系数	coefficient surfacique de déperdition thermique
1.1.5-**3**	bolometer	辐射热测量计	bolomètre
1.1.5-**4**	calorimeter	量热计	calorimètre
1.1.5-**5**	calorimetry	量热学	calorimétrie
1.1.5-**6**	coefficient of expansion	膨胀系数	coefficient de dilatation
1.1.5-**7**	density of heat flow rate *heat flux*	热流密度 热通量	densité de flux thermique *flux thermique surfacique*
1.1.5-**8**	exergetic efficiency *second-law efficiency*	火用效率	rendement exergétique
1.1.5-**9**	heat	热	chaleur
1.1.5-**10**	heat exchanger effectiveness *thermal efficiency (1)*	热效率	efficacité d'un échangeur thermique
1.1.5-**11**	heat flux *density of heat flow rate*	热流密度 热通量	densité de flux thermique *flux thermique surfacique*
1.1.5-**12**	heat gain *heat uptake* ◐	得热量	apport de chaleur *entrée de chaleur* *gain de chaleur*
1.1.5-**13**	heat loss	热损失	perte de chaleur
1.1.5-**14**	heat quantity *amount of heat*	热量	quantité de chaleur
1.1.5-**15**	heat storage *thermal storage*	蓄热	accumulation de chaleur *stockage thermique*
1.1.5-**16**	heat uptake ◐ *heat gain*	得热量	apport de chaleur *entrée de chaleur* *gain de chaleur*
1.1.5-**17**	heating	加热	chauffage *échauffement*
1.1.5-**18**	linear thermal resistance	线性热阻	résistance thermique linéique
1.1.5-**19**	linear thermal transmittance	线性传热率	coefficient linéique de transmission thermique
1.1.5-**20**	overall heat transfer coefficient *thermal transmittance*	对流表面传热系数 总传热系数	coefficient global de transfert de chaleur *coefficient global d'échange thermique* *transmittance thermique* ◐
1.1.5-**21**	reheat (to)	再加热（动词）	réchauffer

ENGLISH	汉语	FRANÇAIS	
reject heat	排热	rejet thermique	1.1.5-**22**
room calorimeter	房间量热计	chambre calorimétrique	1.1.5-**23**
second-law efficiency *exergetic efficiency*	火用效率	rendement exergétique	1.1.5-**24**
sensible heat	显热	chaleur sensible	1.1.5-**25**
thermal conductance	导热	conductance thermique	1.1.5-**26**
thermal diffusivity	热扩散系数 热扩散率	diffusivité thermique	1.1.5-**27**
thermal efficiency (1) *heat exchanger effectiveness*	热效率	efficacité d'un échangeur thermique	1.1.5-**28**
thermal efficiency (2)	热效率	rendement thermodynamique	1.1.5-**29**
thermal effusivity	发射率	effusivité thermique	1.1.5-**30**
(thermal) expansion	热膨胀	dilatation (thermique)	1.1.5-**31**
thermal resistance	热阻	résistance thermique	1.1.5-**32**
thermal storage *heat storage*	蓄热	accumulation de chaleur *stockage thermique*	1.1.5-**33**
thermal transmittance *overall heat transfer coefficient*	对流表面传热系数 总传热系数	coefficient global de transfert de chaleur *coefficient global d'échange thermique* *transmittance thermique* ◐	1.1.5-**34**
volume coefficient of heat loss	容积热损失系数	coefficient volumique de déperdition thermique	1.1.5-**35**
waste heat	废热	chaleur perdue	1.1.5-**36**

SECTION 1.1.6 *Fluid flow*	节 **1.1.6** 流体流动	**SOUS-CHAPITRE 1.1.6** *Ecoulement des fluides*	
absolute viscosity ◐ *dynamic viscosity* *coefficient of viscosity* ◐	动力粘度	viscosité dynamique *coefficient de viscosité* ◐	1.1.6-**1**
airflow	空气流量	débit d'air	1.1.6-**2**
anemometer	风速仪	anémomètre	1.1.6-**3**
annular flow	环流	écoulement annulaire	1.1.6-**4**
aspiration	吸出	aspiration *succion*	1.1.6-**5**
balanced flow	均衡流	flux équilibré	1.1.6-**6**
boundary layer	边界层	couche limite	1.1.6-**7**
bubble flow	泡状流	écoulement à bulles	1.1.6-**8**
buoyancy	浮力	flottabilité	1.1.6-**9**
carrier ring	载流环	bague porteuse	1.1.6-**10**
coefficient of viscosity ◐ *dynamic viscosity* *absolute viscosity* ◐	动力粘度	viscosité dynamique *coefficient de viscosité* ◐	1.1.6-**11**
concentric orifice plate	同心孔板	diaphragme concentrique	1.1.6-**12**
Couette flow	库特流	écoulement de Couette	1.1.6-**13**

	ENGLISH	汉语	FRANÇAIS
1.1.6-**14**	counterflow	逆流	contre-courant
1.1.6-**15**	critical velocity	临界速度	vitesse critique
1.1.6-**16**	cross flow	叉流	écoulements croisés *écoulements transversaux*
1.1.6-**17**	cup anemometer	转杯式风速仪	anémomètre à coupelles
1.1.6-**18**	deflecting vane anemometer	转翼风速仪	anémomètre à palette
1.1.6-**19**	differential pressure device	差压装置	appareil déprimogène
1.1.6-**20**	digital anemometer	数字风速仪	anémomètre à impulsions
1.1.6-**21**	dispersed flow	雾状流	écoulement dispersé
1.1.6-**22**	dynamic loss	动压损失	perte dynamique
1.1.6-**23**	dynamic viscosity *absolute viscosity* ◐ *coefficient of viscosity* ◐	动力粘度	viscosité dynamique *coefficient de viscosité* ◐
1.1.6-**24**	eddy flow ◐ *turbulent flow*	紊流	écoulement turbulent
1.1.6-**25**	electromagnetic flowmeter	电磁流量计	débitmètre électromagnétique
1.1.6-**26**	electronic anemometer	电子风速仪	anémomètre électronique
1.1.6-**27**	emulsion flow	乳状流	écoulement à émulsion *écoulement mousseux* ◐
1.1.6-**28**	face tube ◐ *Pitot tube*	比托管	tube de Pitot
1.1.6-**29**	flow conditioner *flow straightener*	流型调节器	tranquilliseur de débit
1.1.6-**30**	flow nozzle	喷咀	tuyère *ajutage* ◐
1.1.6-**31**	flow pattern	流型	configuration d'écoulement
1.1.6-**32**	flow profile	流速分布图	profil des vitesses
1.1.6-**33**	flow rate (of a fluid) *rate of flow*	流体的流量 流量	débit
1.1.6-**34**	flow stabilizer	稳流器	stabilisateur de débit
1.1.6-**35**	flow straightener *flow conditioner*	流型调节器	tranquilliseur de débit
1.1.6-**36**	flowmeter	流量计	débitmètre
1.1.6-**37**	fluid flow	流体流动	écoulement d'un fluide
1.1.6-**38**	fluidized bed	流态床	lit fluidisé
1.1.6-**39**	friction factor	摩擦系数	coefficient de frottement
1.1.6-**40**	friction loss	摩擦损失	perte de charge frictionnelle *perte de charge par frottement*
1.1.6-**41**	frictional resistance	摩擦阻力	résistance de frottement
1.1.6-**42**	fully developed velocity distribution	充分发展的速度分布	répartition pleinement développée (ou pleinement établie) de la vitesse
1.1.6-**43**	heated thermometer anemometer ◐ *thermal anemometer* *hot-wire anemometer*	热风速仪 热线风速仪	anémomètre thermique *anémomètre à corps chaud* *anémomètre à fil chaud*
1.1.6-**44**	hot-wire anemometer *thermal anemometer* *heated thermometer anemometer* ◐	热风速仪 热线风速仪	anémomètre thermique *anémomètre à corps chaud* *anémomètre à fil chaud*

ENGLISH	汉语	FRANÇAIS	
hydraulic diameter	水力直径	diamètre hydraulique	1.1.6-**45**
interfacial tension ◐ *surface tension*	界面张力 表面张力	tension superficielle *tension interfaciale ◐*	1.1.6-**46**
kinematic viscosity	运动粘度	viscosité cinématique	1.1.6-**47**
lack of miscibility	可混性丧失	lacune de miscibilité *démixtion ◐*	1.1.6-**48**
laminar flow	层流	écoulement laminaire	1.1.6-**49**
mass flow rate	质量流量	débit masse	1.1.6-**50**
mass flux *mass velocity*	质量流速	vitesse massique *densité de flux massique*	1.1.6-**51**
mass velocity *mass flux*	质量流速	vitesse massique *densité de flux massique*	1.1.6-**52**
mean flow rate	平均流量	débit moyen	1.1.6-**53**
mechanical anemometer	机械式风速仪	anémomètre mécanique	1.1.6-**54**
meter tube	测流管	tube de mesurage	1.1.6-**55**
miscibility	可混性	miscibilité	1.1.6-**56**
mist flow	雾流	(écoulement de) buée *(écoulement de) brouillard* *écoulement vésiculaire*	1.1.6-**57**
mixture	混合物	mélange	1.1.6-**58**
molecular flow	分子流	écoulement (en régime) moléculaire *flux moléculaire ◐*	1.1.6-**59**
multi-phase flow	多相流	écoulement multiphasique	1.1.6-**60**
nozzle	喷咀	tuyère	1.1.6-**61**
orifice plate	孔板	diaphragme de mesure *orifice de jaugeage ◐*	1.1.6-**62**
parallel flow	顺流	écoulements parallèles et de même sens *cocourant* *équicourant*	1.1.6-**63**
Pitot tube *face tube ◐*	比托管	tube de Pitot	1.1.6-**64**
plug flow	塞状流	écoulement à bouchons	1.1.6-**65**
porous plug	多孔塞	bouchon poreux	1.1.6-**66**
pulsating flow	脉动流	écoulement pulsatoire	1.1.6-**67**
pulsating flow of mean constant flow rate	平均恒定流量的脉冲流	écoulement pulsatoire à débit moyen constant	1.1.6-**68**
rate of flow *flow rate (of a fluid)*	流体的流量 流量	débit	1.1.6-**69**
regular velocity distribution	正规速度分布	répartition régulière de la vitesse	1.1.6-**70**
revolving vane anemometer	叶轮风速仪	anémomètre à hélice	1.1.6-**71**
Reynolds number (Re)	雷诺数 （Re）	nombre de Reynolds (Re)	1.1.6-**72**
roughness factor	粗糙度	rugosité relative	1.1.6-**73**
slug flow	块状流	***	1.1.6-**74**
sonic velocity *speed of sound*	声速	célérité du son *vitesse du son ◐*	1.1.6-**75**

	ENGLISH	汉语	FRANÇAIS
1.1.6-**76**	speed of sound *sonic velocity*	声速	célérité du son *vitesse du son* ◐
1.1.6-**77**	stagnation point	驻点	point d'arrêt
1.1.6-**78**	stagnation temperature *total temperature* ◐	总温	température d'arrêt *température totale* ◐
1.1.6-**79**	steady flow	稳流	écoulement permanent
1.1.6-**80**	straight length	等截面直管长度	longueur droite
1.1.6-**81**	stratified flow	分层流	écoulement stratifié
1.1.6-**82**	subsonic flow	亚音速流	écoulement subsonique
1.1.6-**83**	supersonic flow	超音速流	écoulement supersonique
1.1.6-**84**	surface tension *interfacial tension* ◐	界面张力 表面张力	tension superficielle *tension interfaciale* ◐
1.1.6-**85**	thermal anemometer *hot-wire anemometer* *heated thermometer anemometer* ◐	热风速仪 热线风速仪	anémomètre thermique *anémomètre à corps chaud* *anémomètre à fil chaud*
1.1.6-**86**	thin orifice plate	薄孔板	diaphragme en mince paroi
1.1.6-**87**	throttling	节流	étranglement
1.1.6-**88**	total temperature ◐ *stagnation temperature*	总温	température d'arrêt *température totale* ◐
1.1.6-**89**	transition flow	过渡流	écoulement de transition
1.1.6-**90**	turbine flowmeter	涡轮流量计	débitmètre à hélice *débitmètre à turbine*
1.1.6-**91**	turbulent flow *eddy flow* ◐	紊流	écoulement turbulent
1.1.6-**92**	two-phase flow	两相流	écoulement diphasique
1.1.6-**93**	ultrasonic flowmeter	超音速流量计	débitmètre à ultrasons
1.1.6-**94**	universal head loss coefficient	通用压头损失系数	coefficient universel de perte de charge
1.1.6-**95**	unsteady flow	不稳定流	écoulement non permanent
1.1.6-**96**	vane anemometer	叶轮式风速仪	anémomètre à moulinet
1.1.6-**97**	velocity distribution	速度分布	répartition des vitesses
1.1.6-**98**	velocity of flow	流速	vitesse d'écoulement
1.1.6-**99**	velocity profile	流速分布图	profil des vitesses
1.1.6-**100**	Venturi tube	文丘里管	tube de Venturi
1.1.6-**101**	viscometer *viscosimeter*	粘度计	viscosimètre
1.1.6-**102**	viscosimeter *viscometer*	粘度计	viscosimètre
1.1.6-**103**	viscosity	粘度	viscosité
1.1.6-**104**	viscous flow	粘性流	écoulement visqueux
1.1.6-**105**	volume flow rate *volumetric flow rate*	体积流量	débit volume
1.1.6-**106**	volumetric flow rate *volume flow rate*	体积流量	débit volume
1.1.6-**107**	vortex flowmeter	涡流流量计	débitmètre à vortex
1.1.6-**108**	wavy flow	波状流	écoulement ondulé

ENGLISH	汉语	FRANÇAIS	
SECTION 1.1.7 *Metrology – Measuring apparatus*	节 **1.1.7** 测量学——测量装置	**SOUS-CHAPITRE 1.1.7** *Métrologie – Appareils de mesure*	
accuracy	准确度	exactitude	1.1.7-**1**
accuracy of measurement	测量精度	exactitude de mesure	1.1.7-**2**
adjustment (of a measuring instrument)	（测量仪器的）校正	ajustage (d'un instrument de mesure)	1.1.7-**3**
analogue indicating instrument *analogue measuring instrument*	模拟显示仪	appareil de mesure (à affichage) analogique	1.1.7-**4**
analogue measuring instrument *analogue indicating instrument*	模拟测量仪	appareil de mesure (à affichage) analogique	1.1.7-**5**
base quantity	基量	grandeur de base	1.1.7-**6**
calibration	标定	étalonnage	1.1.7-**7**
case	（压力计）外壳	boîtier	1.1.7-**8**
certified standard instrument	校正的标准仪表	instrument étalonné agréé	1.1.7-**9**
chart	图表	support d'enregistrement	1.1.7-**10**
controlled variable	受控变量	grandeur réglée	1.1.7-**11**
correcting variable	校正变量	grandeur réglante	1.1.7-**12**
derived quantity	导出量	grandeur dérivée	1.1.7-**13**
detecting element *sensor* *sensing element* ◐	传感器 检测元件 敏感元件	capteur *élément sensible*	1.1.7-**14**
detector	检测器	détecteur	1.1.7-**15**
dial	刻度盘	cadran	1.1.7-**16**
digital indicating instrument *digital measuring instrument*	数字显示仪	appareil de mesure (à affichage) numérique	1.1.7-**17**
digital measuring instrument *digital indicating instrument*	数字检测仪	appareil de mesure (à affichage) numérique	1.1.7-**18**
dimensionless quantity *quantity of dimension one* ○	无因次量	grandeur adimensionnelle *grandeur de dimension un*	1.1.7-**19**
displaying device *indicating device*	显示装置	dispositif d'affichage *dispositif indicateur*	1.1.7-**20**
displaying (measuring) instrument *indicating (measuring) instrument*	显示（测量）仪表	appareil (de mesure) afficheur *appareil (de mesure) indicateur*	1.1.7-**21**
error of measurement	测量误差	erreur de mesure *erreur de mesurage*	1.1.7-**22**
flange	法兰	collerette	1.1.7-**23**
gauging (of a measuring instrument)	（测量仪表的）刻度	calibrage (d'un instrument de mesure)	1.1.7-**24**
indicating device *displaying device*	显示装置	dispositif d'affichage *dispositif indicateur*	1.1.7-**25**
indicating (measuring) instrument *displaying (measuring) instrument*	显示（测量）仪表	appareil (de mesure) afficheur *appareil (de mesure) indicateur*	1.1.7-**26**
indication (of a measuring instrument)	（测量仪表的）显示	indication (d'un instrument de mesure)	1.1.7-**27**
influence quantity	干扰量	grandeur d'influence	1.1.7-**28**
input (variable)	输入（变量）	grandeur d'entrée	1.1.7-**29**
integrating (measuring) instrument	积分（测量）仪表	appareil (de mesure) intégrateur	1.1.7-**30**

	ENGLISH	汉语	FRANÇAIS
1.1.7-**31**	International System of Units (SI)	国际单位制 （SI）	Système International d'unités (SI)
1.1.7-**32**	intrinsic error (of a measuring instrument)	（测量仪表的）内在误差	erreur intrinsèque (d'un instrument de mesure)
1.1.7-**33**	limiting conditions	极限条件	conditions limites
1.1.7-**34**	(measurable) quantity	可测量	grandeur (mesurable)
1.1.7-**35**	measurand	被测量	mesurande
1.1.7-**36**	measurement	测量	mesurage
1.1.7-**37**	measurement procedure	测量步骤	mode opératoire (du mesurage)
1.1.7-**38**	measurement value (of a quantity) *result of a measurement*	测量值 测量结果	mesure *résultat du mesurage*
1.1.7-**39**	measuring instrument	测量仪表	instrument de mesure *appareil de mesure*
1.1.7-**40**	measuring range *working range*	测量范围 工作范围	étendue de mesure *plage de mesure*
1.1.7-**41**	measuring transducer	测量变送器	transducteur de mesurage *transducteur de mesure* *jauge de mesurage*
1.1.7-**42**	method of measurement	测量方法	méthode de mesure
1.1.7-**43**	metrology	测量学	métrologie
1.1.7-**44**	nominal range	名义测量范围	calibre *plage*
1.1.7-**45**	output (variable)	输出(变量)	grandeur de sortie
1.1.7-**46**	precision	精确度	précision
1.1.7-**47**	pressure sensor	压力传感器	capteur de pression *jauge de pression*
1.1.7-**48**	quantity of dimension one ○ *dimensionless quantity*	无因次量	grandeur adimensionnelle *grandeur de dimension un*
1.1.7-**49**	random error	随机误差	erreur aléatoire
1.1.7-**50**	range of indication	指示范围	étendue des indications *étendue d'échelle*
1.1.7-**51**	rated operating conditions	标定工况	conditions assignées de fonctionnement
1.1.7-**52**	recording device	记录装置	dispositif enregistreur
1.1.7-**53**	recording duration	记录持续时间	durée d'enregistrement
1.1.7-**54**	recording interval	记录时间间隔	période d'enregistrement *intervalle d'enregistrement*
1.1.7-**55**	recording (measuring) instrument	记录(测量)仪表	appareil (de mesure) enregistreur
1.1.7-**56**	relative error	相对误差	erreur relative
1.1.7-**57**	repeatability of measurements	测量的重复性	répétabilité des mesurages
1.1.7-**57**	reproducibility of measurements	测量的再现性	reproductibilité des mesurages
1.1.7-**59**	resolution	分辨度	résolution
1.1.7-**60**	response time	响应时间	temps de réponse
1.1.7-**61**	result of a measurement *measurement value (of a quantity)*	测量值 测量结果	résultat du mesurage *mesure*
1.1.7-**62**	scale division	刻度盘的分度	division
1.1.7-**63**	scale length	刻度长度	longueur d'échelle

ENGLISH	汉语	FRANÇAIS	
scale (of a measuring instrument)	（测量仪表的）刻度	échelle (d'un instrument de mesure)	1.1.7-**64**
sensitivity	灵敏度	sensibilité	1.1.7-**65**
sensing element ◐ *sensor* *detecting element*	传感器 检测元件 敏感元件	capteur *élément sensible*	1.1.7-**66**
sensor *detecting element* *sensing element* ◐	传感器 检测元件 敏感元件	capteur *élément sensible*	1.1.7-**67**
span	量程	intervalle de mesure	1.1.7-**68**
stability	稳定性	constance	1.1.7-**69**
storage and transport conditions	储藏和运输条件	conditions de stockage et de transport	1.1.7-**70**
symbol of unit (of measurement)	（测量）单位符号	symbole d'une unité (de mesure)	1.1.7-**71**
temperature sensor	温度传感器	capteur de température	1.1.7-**72**
transducer	变送器	transducteur	1.1.7-**73**
uncertainty of measurement	测量的不确定性	incertitude de mesure	1.1.7-**74**
unit (of measurement)	（测量）单位	unité (de mesure)	1.1.7-**75**
user adjustment (of a measuring instrument)	（测量仪表的）用户校正	réglage (d'un instrument de mesure)	1.1.7-**76**
value (of a quantity)	（数量）值	valeur (d'une grandeur)	1.1.7-**77**
working range *measuring range*	测量范围 工作范围	étendue de mesure *plage de mesure*	1.1.7-**78**

SECTION 1.2 *Thermodynamic properties and processes* **SECTION 1.2.1** *Thermodynamic properties and processes: general background*	节 **1.2** 热力性质及过程 节 **1.2.1** 热力性质及过程：基础知识	**SOUS-CHAPITRE 1.2** *Propriétés et transformations thermodynamiques* **SOUS-CHAPITRE 1.2.1** *Généralités sur les propriétés et les transformations thermodynamiques*	
adiabatic	绝热的	adiabatique	1.2.1-**1**
availability *exergy*	火用	exergie	1.2.1-**2**
availability destruction	火用损	perte d'exergie *destruction d'exergie*	1.2.1-**3**
boundary conditions	边界条件	conditions aux limites	1.2.1-**4**
closed process	闭合过程	système fermé	1.2.1-**5**
coefficient of compressibility ◐ *compressibility factor*	压缩因子	facteur de compressibilité	1.2.1-**6**
cold reservoir *low-temperature reservoir* *heat source (2)* ◐	低温热源 低温热库 低温冷库	source froide (absorbant la chaleur)	1.2.1-**7**
cold source ○ *heat sink*	热汇	puits de chaleur *source de froid (au sens banal)* ◐	1.2.1-**8**
compressibility factor *coefficient of compressibility* ◐	压缩因子	facteur de compressibilité	1.2.1-**9**
corresponding state(s)	对应态	états correspondants	1.2.1-**10**

	ENGLISH	汉语	FRANÇAIS
1.2.1-**11**	cyclic thermal conditions ◐	循环热工况	régime cyclique *régime variable*
1.2.1-**12**	energy breakdown (in exergetic sense) ◐	能量品位降低	dégradation de l'énergie (dans le sens d'exergie)
1.2.1-**13**	energy conservation (1)	能量守恒	conservation de l'énergie
1.2.1-**14**	energy conservation (2) *first law (of thermodynamics)* *equivalence principle* ◐ *first principle (of thermodynamics)* ◐	热力学第一定律	premier principe (de la thermodynamique) *principe de l'équivalence* ◐
1.2.1-**15**	energy level *energy state*	能级	niveau d'énergie
1.2.1-**16**	energy state *energy level*	能级	niveau d'énergie
1.2.1-**17**	enthalpy	焓	enthalpie
1.2.1-**18**	entropy	熵	entropie
1.2.1-**19**	entropy generation *entropy production*	熵产	création d'entropie
1.2.1-**20**	entropy production *entropy generation*	熵产	création d'entropie
1.2.1-**21**	equivalence principle ◐ *first law (of thermodynamics)* *energy conservation (2)* *first principle (of thermodynamics)* ◐	热力学第一定律	premier principe (de la thermodynamique) *principe de l'équivalence* ◐
1.2.1-**22**	exergetic efficiency *second-law efficiency*	火用效率	rendement exergétique
1.2.1-**23**	exergy *availability*	火用	exergie
1.2.1-**24**	first law (of thermodynamics) *energy conservation (2)* *equivalence principle* ◐ *first principle (of thermodynamics)* ◐	热力学第一定律	premier principe (de la thermodynamique) *principe de l'équivalence* ◐
1.2.1-**25**	first principle (of thermodynamics) ◐ *first law (of thermodynamics)* *energy conservation (2)* *equivalence principle* ◐	热力学第一定律	premier principe (de la thermodynamique) *principe de l'équivalence* ◐
1.2.1-**26**	fugacity	逸度	fugacité
1.2.1-**27**	heat equivalent of work ○	功热当量	équivalent calorifique de travail ○
1.2.1-**28**	heat reservoir (1)	热库	source de chaleur (au sens thermo-dynamique)
1.2.1-**29**	heat reservoir (2) *high-temperature reservoir* *hot reservoir*	高温热库	source chaude
1.2.1-**30**	heat sink *cold source* ○	热汇	puits de chaleur *source de froid (au sens banal)* ◐
1.2.1-**31**	heat source (1)	热源	source de chaleur (au sens banal)
1.2.1-**32**	heat source (2) *cold reservoir* *low-temperature reservoir*	低温热源 低温热库 低温冷库	source froide (absorbant la chaleur)
1.2.1-**33**	high-temperature reservoir *heat reservoir (2)* *hot reservoir*	高温热库	source chaude

ENGLISH	汉语	FRANÇAIS	
hot reservoir *heat reservoir (2)* *high-temperature reservoir*	高温热库	source chaude	1.2.1-**34**
ideal gas	理想气体	gaz idéal	1.2.1-**35**
(ideal) gas constant *(perfect) gas constant* ◐	理想气体常数	constante des gaz idéaux (ou idéals) *constante des gaz parfaits*	1.2.1-**36**
(ideal) gas equation *(perfect) gas equation* ◐	理想气体方程	équation des gaz idéaux (ou idéals) *équation des gaz parfaits*	1.2.1-**37**
ideal multistage compression	理想多级压缩	compression polyétagée idéale	1.2.1-**38**
internal energy *intrinsic energy* ◐	内能	énergie interne	1.2.1-**39**
intrinsic energy ◐ *internal energy*	内能	énergie interne	1.2.1-**40**
irreversibility	不可逆性	irréversibilité	1.2.1-**41**
irreversible process	不可逆过程	transformation irréversible	1.2.1-**42**
isolated system	孤立系统	système isolé	1.2.1-**43**
law of corresponding states	对比态定律	loi des états correspondants	1.2.1-**44**
low-temperature reservoir *cold reservoir* *heat source (2)* ◐	低温热源 低温热库 低温冷库	source froide (absorbant la chaleur)	1.2.1-**45**
mechanical equivalent of heat ○	热功当量	équivalent mécanique de chaleur ○	1.2.1-**46**
nonequilibrium thermodynamics	不平衡热力学	thermodynamique des processus irréversibles	1.2.1-**47**
open process	开口过程	système ouvert	1.2.1-**48**
perfect gas	理想气体	gaz parfait	1.2.1-**49**
(perfect) gas constant ◐ *(ideal) gas constant*	理想气体常数	constante des gaz idéaux (ou idéals) *constante des gaz parfaits*	1.2.1-**50**
(perfect) gas equation ◐ *(ideal) gas equation*	理想气体方程	équation des gaz idéaux (ou idéals) *équation des gaz parfaits*	1.2.1-**51**
permanent thermal conditions *steady thermal conditions*	稳态热工况	régime (thermique) permanent *régime (thermique) stationnaire* ◐	1.2.1-**52**
real fluid	真实流体	fluide réel	1.2.1-**53**
real gas	真实气体	gaz réel	1.2.1-**54**
reversible process	可逆过程	transformation réversible	1.2.1-**55**
second law (of thermodynamics) *second principle (of thermodynamics)* ◐	热力学第二定律	second principe (de la thermodynamique)	1.2.1-**56**
second-law efficiency *exergetic efficiency*	火用效率	rendement exergétique	1.2.1-**57**
second principle (of thermodynamics) ◐ *second law (of thermodynamics)*	热力学第二定律	second principe (de la thermodynamique)	1.2.1-**58**
stationary state ◐ *steady state*	稳态	régime permanent *régime établi* ◐ *régime stable* ◐	1.2.1-**59**
statistical thermodynamics	统计热力学	thermodynamique statistique	1.2.1-**60**
steady state *stationary state* ◐	稳态	régime permanent *régime établi* ◐ *régime stable* ◐	1.2.1-**61**
steady thermal conditions *permanent thermal conditions*	稳态热工况	régime (thermique) permanent *régime (thermique) stationnaire* ◐	1.2.1-**62**

	ENGLISH	汉语	FRANÇAIS
1.2.1-**63**	surroundings	环境	milieu extérieur
1.2.1-**64**	(thermal) equation of state	状态方程式	équation d'état (thermique)
1.2.1-**65**	thermal equilibrium	热平衡	équilibre thermique
1.2.1-**66**	thermodynamic equilibrium	热力平衡	équilibre thermodynamique
1.2.1-**67**	thermodynamic similarity	热力相似	similitude thermodynamique
1.2.1-**68**	(thermodynamic) system	热力学系统	système (thermodynamique)
1.2.1-**69**	thermodynamics	热力学	thermodynamique
1.2.1-**70**	thermophysics	热物理	thermophysique
1.2.1-**71**	third law (of thermodynamics) *third principle (of thermodynamics)* ◐	热力学第三定律	troisième principe (de la thermodynamique)
1.2.1-**72**	third principle (of thermodynamics) ◐ *third law (of thermodynamics)*	热力学第三定律	troisième principe (de la thermodynamique)
1.2.1-**73**	transient state	瞬态	régime transitoire
1.2.1-**74**	transport property	迁移特性	propriété de transport
1.2.1-**75**	unsteady state	非稳态	régime variable
1.2.1-**76**	virial coefficients	维里系数	coefficients du viriel
1.2.1-**77**	zero point energy	零点能	énergie de point zéro
1.2.1-**78**	zero principle *zeroth law of thermodynamics*	零定理 热力学零定理	principe zéro
1.2.1-**79**	zeroth law of thermodynamics *zero principle*	零定理 热力学零定理	principe zéro

	SECTION 1.2.2 *Phase change*	节 **1.2.2** 相变	**SOUS-CHAPITRE 1.2.2** *Changement de phase*
1.2.2-**1**	atmospheric boiling point *normal boiling point*	标准沸点	point d'ébullition *température d'ébullition*
1.2.2-**2**	azeotrope *azeotropic mixture*	共沸混合物	azéotrope *mélange azéotropique* ◐
1.2.2-**3**	azeotropic	共沸的	azéotrope (adj.) *azéotropique*
1.2.2-**4**	azeotropic mixture *azeotrope*	共沸混合物	azéotrope *mélange azéotropique* ◐
1.2.2-**5**	azeotropic point	共沸点	point azéotropique *point d'azéotropie* ◐
1.2.2-**6**	azeotropy	共沸性	azéotropie
1.2.2-**7**	boiling *ebullition* ◐	沸腾	ébullition
1.2.2-**8**	bubble point	始沸点	point de bulle
1.2.2-**9**	burnout point *critical (nucleate boiling) heat flux* *maximum nucleate boiling heat flux* *peak nucleate boiling heat flux*	核状沸腾最大热通量 核状沸腾临界热通量 核状沸腾峰值热通量	flux maximal de l'ébullition nucléée *flux critique de l'ébullition nucléée*
1.2.2-**10**	change of state (1)	相变	changement d'état
1.2.2-**11**	change of state (2) ◐ *phase change*	相变	changement de phase *transition de phase*

ENGLISH	汉语	FRANÇAIS	
chemical potential	化学势	potentiel chimique	1.2.2-**12**
condensate	冷凝液	condensat *produit de condensation* ◐	1.2.2-**13**
condensation *liquefaction*	冷凝 液化	condensation *liquéfaction* ◐	1.2.2-**14**
condensation point	凝结点	point de condensation	1.2.2-**15**
critical (nucleate boiling) heat flux *burnout point* *maximum nucleate boiling heat flux* *peak nucleate boiling heat flux*	核状沸腾最大热通量 核状沸腾临界热通量 核状沸腾峰值热通量	flux maximal de l'ébullition nucléée *flux critique de l'ébullition nucléée*	1.2.2-**16**
critical point *critical state*	临界点 临界状态	état critique	1.2.2-**17**
critical pressure	临界压力	pression critique	1.2.2-**18**
critical state *critical point*	临界点 临界状态	état critique	1.2.2-**19**
critical temperature	临界温度	température critique	1.2.2-**20**
critical volume	临界比体积	volume critique	1.2.2-**21**
cryoscopy	冰点测定法	cryoscopie *cryométrie*	1.2.2-**22**
depression of freezing point	冻结点差值	abaissement du point de congélation	1.2.2-**23**
dew point	露点	point de rosée	1.2.2-**24**
droplet condensation	滴状凝结	condensation en gouttes	1.2.2-**25**
(dry) saturated vapour	(干)饱和蒸汽	vapeur saturante *vapeur saturée (sèche)*	1.2.2-**26**
ebullition ◐ *boiling*	沸腾	ébullition	1.2.2-**27**
eutectic	低共熔混合物	eutectique	1.2.2-**28**
eutectic point	低共熔点	point d'eutexie *point eutectique* ◐	1.2.2-**29**
eutexy	低共熔性	eutexie	1.2.2-**30**
evaporation	蒸发	évaporation	1.2.2-**31**
evaporation rate	蒸发速率	débit d'évaporation	1.2.2-**32**
film boiling	膜状沸腾	ébullition en film *ébullition pelliculaire* ◐	1.2.2-**33**
film condensation	膜状凝结	condensation en film *condensation pelliculaire* ◐	1.2.2-**34**
fluid	流体	fluide	1.2.2-**35**
freezing	冻结	congélation	1.2.2-**36**
freezing point	冻结点	point de congélation *point de solidification* ◐	1.2.2-**37**
ice melting equivalent	冰融当量	chaleur latente de la glace	1.2.2-**38**
ice (melting) point	冰(的融)点	point de (fusion de la) glace *température de la glace fondante* ◐	1.2.2-**39**
latent heat	潜热	chaleur latente	1.2.2-**40**
liquefaction *condensation*	冷凝 液化	condensation *liquéfaction* ◐	1.2.2-**41**
liquefaction of gases	气体液化	liquéfaction des gaz	1.2.2-**42**

	ENGLISH	汉语	FRANÇAIS
1.2.2-**43**	liquid-vapour mixture	汽-液混合物	mélange liquide-vapeur
1.2.2-**44**	maximum nucleate boiling heat flux *burnout point* *critical (nucleate boiling) heat flux* *peak nucleate boiling heat flux*	核状沸腾最大热通量 核状沸腾临界热通量 核状沸腾峰值热通量	flux maximal de l'ébullition nucléée *flux critique de l'ébullition nucléée*
1.2.2-**45**	melting	熔化	fusion
1.2.2-**46**	melting point	融点	point de fusion
1.2.2-**47**	non-azeotropic mixture *zeotrope* *zeotropic mixture*	非共沸混合物	zéotrope *mélange non-azéotropique* ◐ *mélange zéotropique* ◐
1.2.2-**48**	normal boiling point *atmospheric boiling point*	标准沸点	point d'ébullition *température d'ébullition*
1.2.2-**49**	nucleate boiling	核状沸腾	ébullition nucléée
1.2.2-**50**	nucleation	成核现象	nucléation
1.2.2-**51**	peak nucleate boiling heat flux *burnout point* *critical (nucleate boiling) heat flux* *maximum nucleate boiling heat flux*	核状沸腾最大热通量 核状沸腾临界热通量 核状沸腾峰值热通量	flux maximal de l'ébullition nucléée *flux critique de l'ébullition nucléée*
1.2.2-**52**	phase change *change of state (2)* ◐	相变	changement de phase *transition de phase*
1.2.2-**53**	pool boiling	池沸腾	ébullition libre
1.2.2-**54**	quality	干度	titre en vapeur
1.2.2-**55**	saturated liquid	饱和液体	liquide saturant
1.2.2-**56**	saturated vapour pressure	饱和蒸汽压	pression de vapeur saturante
1.2.2-**57**	saturation	饱和	saturation
1.2.2-**58**	saturation pressure	饱和压力	pression de saturation
1.2.2-**59**	saturation temperature	饱和温度	température de saturation
1.2.2-**60**	solidification	凝固	solidification
1.2.2-**61**	solidification point	凝固点	point de solidification
1.2.2-**62**	subcooled liquid	过冷液体	liquide sous-refroidi
1.2.2-**63**	subcooling	过冷却	sous-refroidissement
1.2.2-**64**	sublimation (desublimation)	升华	sublimation (désublimation)
1.2.2-**65**	(suction) superheat	吸气过热	surchauffe à l'aspiration
1.2.2-**66**	supercooling	过冷	surfusion
1.2.2-**67**	superheat	过热	surchauffe
1.2.2-**68**	superheat (to)	过热	surchauffer
1.2.2-**69**	superheated vapour	过热蒸汽	vapeur surchauffée
1.2.2-**70**	supersaturated vapour	过饱和蒸汽	vapeur sursaturée
1.2.2-**71**	supersaturation	过饱和	sursaturation
1.2.2-**72**	triple point	三相点	point triple
1.2.2-**73**	vaporization	汽化	vaporisation
1.2.2-**74**	vapour	蒸汽	vapeur
1.2.2-**75**	wet vapour	湿蒸汽	vapeur humide

ENGLISH	汉语	FRANÇAIS	
zeotrope *non-azeotropic mixture* *zeotropic mixture*	非共沸混合物	zéotrope *mélange non-azéotropique* ◐ *mélange zéotropique* ◐	1.2.2-**76**
zeotropic mixture *non-azeotropic mixture* *zeotrope*	非共沸混合物	zéotrope *mélange non-azéotropique* ◐ *mélange zéotropique* ◐	1.2.2-**77**

SECTION 1.2.3 *Thermodynamic processes*	节 **1.2.3** 热力过程	**SOUS-CHAPITRE 1.2.3** *Transformations thermodynamiques*	
absorption (refrigeration) cycle	吸收式(制冷)循环	cycle (frigorifique) à absorption	1.2.3-**1**
actual cycle	实际循环	cycle réel	1.2.3-**2**
adiabatic	绝热的	adiabatique	1.2.3-**3**
adiabatic compression	绝热压缩	compression adiabatique	1.2.3-**4**
adiabatic expansion	绝热膨胀	détente adiabatique	1.2.3-**5**
adiabatic exponent	绝热指数	exposant adiabatique	1.2.3-**6**
air (refrigeration) cycle	空气(制冷)循环	cycle (frigorifique) à air	1.2.3-**7**
Carnot cycle	卡诺循环	cycle de Carnot	1.2.3-**8**
closed cycle	闭式循环	cycle en circuit "fermé"	1.2.3-**9**
compressibility	可压缩性	compressibilité	1.2.3-**10**
compression	压缩	compression	1.2.3-**11**
compression exponent *compression index*	压缩指数	exposant de compression	1.2.3-**12**
compression index *compression exponent*	压缩指数	exposant de compression	1.2.3-**13**
compression (refrigeration) cycle	压缩(制冷)循环	cycle (frigorifique) à compression	1.2.3-**14**
degrees of freedom *degrees of liberty* ○	自由度	degrés de liberté *variance*	1.2.3-**15**
degrees of liberty ○ *degrees of freedom*	自由度	degrés de liberté *variance*	1.2.3-**16**
enthalpy chart *enthalpy diagram*	焓图	diagramme enthalpique	1.2.3-**17**
enthalpy diagram *enthalpy chart*	焓图	diagramme enthalpique	1.2.3-**18**
entropy chart *entropy diagram*	熵图	diagramme entropique	1.2.3-**19**
entropy diagram *entropy chart*	熵图	diagramme entropique	1.2.3-**20**
expander cycle *work cycle* *work extraction cycle*	膨胀机循环 做功循环	cycle avec travail extérieur	1.2.3-**21**
expansion	膨胀	détente	1.2.3-**22**
indicator diagram	示功图	diagramme indicateur	1.2.3-**23**
isenthalp	等焓线	isenthalpe	1.2.3-**24**
isenthalpic	等焓的	isenthalpique	1.2.3-**25**
isenthalpic expansion	等焓膨胀	détente isenthalpique	1.2.3-**26**

	ENGLISH	汉语	FRANÇAIS
1.2.3-**27**	isentrope	等熵线	isentrope
1.2.3-**28**	isentropic	等熵的	isentropique
1.2.3-**29**	isobar	等压线	isobare (subst.)
1.2.3-**30**	isobaric	等压的	isobare (adj.)
1.2.3-**31**	isochor	等容线	isochore (subst.)
1.2.3-**32**	isochoric	等容的	isochore (adj.) *isovolume* ◐
1.2.3-**33**	isotherm	等温线	isotherme (subst.)
1.2.3-**34**	isothermal	等温的	isotherme (adj.) *isothermique* ◐
1.2.3-**35**	(isothermal) compressibility	（等温）压缩系数	compressibilité (isotherme)
1.2.3-**36**	Joule-Thomson effect	焦耳-汤姆逊效应	effet Joule-Thomson
1.2.3-**37**	Joule-Thomson process	焦耳-汤姆逊过程	détente sans travail extérieur
1.2.3-**38**	Lorenz cycle	洛伦兹循环	cycle de Lorenz
1.2.3-**39**	Mollier chart *Mollier diagram*	莫利尔图	diagramme de Mollier
1.2.3-**40**	Mollier diagram *Mollier chart*	莫利尔图	diagramme de Mollier
1.2.3-**41**	open cycle	开式循环	cycle ouvert
1.2.3-**42**	polytrop	多变过程线	polytrope
1.2.3-**43**	polytropic	多变的	polytropique
1.2.3-**44**	polytropic compression	多变压缩	compression polytropique
1.2.3-**45**	polytropic expansion	多变膨胀	détente polytropique
1.2.3-**46**	polytropic exponent	多变指数	exposant polytropique
1.2.3-**47**	polytropic process	多变过程	processus polytropique
1.2.3-**48**	pressure volume chart *PV chart* *pressure volume diagram*	压容图	diagramme de Clapeyron *diagramme pression volume* *diagramme PV* ◐
1.2.3-**49**	pressure volume diagram *PV chart* *pressure volume chart*	压容图	diagramme de Clapeyron *diagramme pression volume* *diagramme PV* ◐
1.2.3-**50**	PV chart *pressure volume chart* *pressure volume diagram*	压容图	diagramme de Clapeyron *diagramme pression volume* *diagramme PV* ◐
1.2.3-**51**	Rankine cycle	朗肯循环	cycle de Rankine
1.2.3-**52**	refrigeration cycle	制冷循环	cycle frigorifique
1.2.3-**53**	reverse cycle	逆循环	cycle inversé
1.2.3-**54**	reversible cycle	可逆循环	cycle réversible
1.2.3-**55**	semi-closed cycle	半闭式循环	cycle semi-fermé
1.2.3-**56**	standard rating cycle	标准工况循环	cycle de référence
1.2.3-**57**	state diagram	状态图	diagramme d'état
1.2.3-**58**	steam jet (refrigeration) cycle *vapour jet (refrigeration) cycle*	蒸汽喷射（制冷）循环	cycle (frigorifique) à éjection (de vapeur)
1.2.3-**59**	Stirling cycle	斯特林循环	cycle de Stirling
1.2.3-**60**	(thermodynamic) cycle	（热力）循环	cycle (thermodynamique)

ENGLISH	汉语	FRANÇAIS	
throttling expansion	节流膨胀	détente sans travail extérieur	1.2.3-**61**
vapour jet (refrigeration) cycle *steam jet (refrigeration) cycle*	蒸汽喷射(制冷)循环	cycle (frigorifique) à éjection (de vapeur)	1.2.3-**62**
work cycle *expander cycle* *work extraction cycle*	膨胀机循环 做功循环	cycle avec travail extérieur	1.2.3-**63**
work extraction cycle *expander cycle* *work cycle*	膨胀机循环 做功循环	cycle avec travail extérieur	1.2.3-**64**
working fluid *working substance*	工作流体 工质	fluide actif	1.2.3-**65**
working substance *working fluid*	工作流体 工质	fluide actif	1.2.3-**66**

SECTION 1.3 *Heat and mass transfer* **SECTION 1.3.1** *Heat and mass transfer: general background*	节 **1.3** 传热与传质 节 **1.3.1** 传热与传质：基础知识	**SOUS-CHAPITRE 1.3** *Transfert de chaleur et de masse* **SOUS-CHAPITRE 1.3.1** *Généralités sur le transfert de chaleur et de masse*	
cooling range	冷却范围	amplitude du refroidissement	1.3.1-**1**
energy flow rate	能流速率	flux énergétique	1.3.1-**2**
film coefficient of heat transfer ◐ *heat transfer coefficient* *surface coefficient of heat transfer* ◐ *surface film conductance* ◐	对流换热表面传热系数 表面传热系数	coefficient de transfert de chaleur *coefficient d'échange (thermique) superficiel* *coefficient de transmission (thermique) de surface*	1.3.1-**3**
heat	热	chaleur	1.3.1-**4**
heat exchange	换热	échange thermique *échange de chaleur* ◐	1.3.1-**5**
heat flow	热流	transport thermique *transport de chaleur* ◐	1.3.1-**6**
heat flow meter	热流计	fluxmètre thermique	1.3.1-**7**
heat flow path	热流线路	ligne de flux thermique *ligne de flux de chaleur* ◐	1.3.1-**8**
heat flow rate	热流率	flux thermique	1.3.1-**9**
heat lag ◐ *thermal lag*	热迟延	temps mort thermique	1.3.1-**10**
heat transfer *heat transmission* ◐ *heat transport* ◐ *thermal transmission* ◐	传热	transfert de chaleur *transmission de chaleur* ◐	1.3.1-**11**
heat transfer coefficient *film coefficient of heat transfer* ◐ *surface coefficient of heat transfer* ◐ *surface film conductance* ◐	对流换热表面传热系数 表面传热系数	coefficient de transfert de chaleur *coefficient d'échange (thermique) superficiel* *coefficient de transmission (thermique) de surface*	1.3.1-**12**
heat transmission ◐ *heat transfer* *heat transport* ◐ *thermal transmission* ◐	传热	transfert de chaleur *transmission de chaleur* ◐	1.3.1-**13**

	ENGLISH	汉语	FRANÇAIS
1.3.1-**14**	heat transport ◐ *heat transfer* *heat transmission* ◐ *thermal transmission* ◐	传热	transfert de chaleur *transmission de chaleur* ◐
1.3.1-**15**	overall heat transfer coefficient *thermal transmittance*	对流表面传热系数 总传热系数	coefficient global de transfert de chaleur *coefficient global d'échange thermique* *transmittance thermique* ◐
1.3.1-**16**	surface coefficient of heat transfer ◐ *heat transfer coefficient* *film coefficient of heat transfer* ◐ *surface film conductance* ◐	对流换热表面传热系数 表面传热系数	coefficient de transfert de chaleur *coefficient d'échange (thermique) superficie* *coefficient de transmission (thermique)* *de surface*
1.3.1-**17**	surface film conductance ◐ *heat transfer coefficient* *film coefficient of heat transfer* ◐ *surface coefficient of heat transfer* ◐	对流换热表面传热系数 表面传热系数	coefficient de transfert de chaleur *coefficient d'échange (thermique) superficie* *coefficient de transmission (thermique)* *de surface*
1.3.1-**18**	temperature field	温度场	champ de température *champ thermique*
1.3.1-**19**	thermal boundary resistance *thermal contact resistance*	接触热阻	résistance thermique de contact *résistance thermique limite* ◐
1.3.1-**20**	thermal contact resistance *thermal boundary resistance*	接触热阻	résistance thermique de contact *résistance thermique limite* ◐
1.3.1-**21**	thermal lag *heat lag* ◐	热迟延	temps mort thermique
1.3.1-**22**	thermal resistance	热阻	résistance thermique
1.3.1-**23**	thermal transmission ◐ *heat transfer* *heat transmission* ◐ *heat transport* ◐	传热	transfert de chaleur *transmission de chaleur* ◐
1.3.1-**24**	thermal transmittance *overall heat transfer coefficient*	对流表面传热系数 总传热系数	coefficient global de transfert de chaleur *coefficient global d'échange thermique* *transmittance thermique* ◐
1.3.1-**25**	transient heat flow	瞬变热流	écoulement de chaleur en régime transitoire
	SECTION 1.3.2 *Radiation*	节 **1.3.2** 辐射	**SOUS-CHAPITRE 1.3.2** *Rayonnement*
1.3.2-**1**	absorptance *absorptivity* *absorption factor* ◐	吸收率	absorptance *absorptivité* *facteur d'absorption*
1.3.2-**2**	absorption coefficient	吸收系数	coefficient d'absorption
1.3.2-**3**	absorption factor ◐ *absorptance* *absorptivity*	吸收率	absorptance *absorptivité* *facteur d'absorption*
1.3.2-**4**	absorptivity *absorptance* *absorption factor* ◐	吸收率	absorptance *absorptivité* *facteur d'absorption*
1.3.2-**5**	angle factor *shape factor* *view factor* *configuration factor* ◐	角系数 形状系数 视角因数	facteur d'angle *facteur de forme*
1.3.2-**6**	athermanous	不透辐射热的	athermane

ENGLISH	汉语	FRANÇAIS	
black body	黑体	corps noir	1.3.2-**7**
configuration factor ◐ *angle factor* *shape factor* *view factor*	角系数 形状系数 视角因数	facteur d'angle *facteur de forme*	1.3.2-**8**
diathermanous	透辐射热的	diathermane	1.3.2-**9**
diffuse surface	漫射表面	surface diffusante	1.3.2-**10**
emissive power *emittance* *radiant exitance*	辐射强度	émittance *exitance*	1.3.2-**11**
emissivity	辐射率	émissivité	1.3.2-**12**
emittance *emissive power* *radiant exitance*	辐射强度	émittance *exitance*	1.3.2-**13**
gray body (USA) *grey body*	灰体	corps gris	1.3.2-**14**
grey body *gray body (USA)*	灰体	corps gris	1.3.2-**15**
interstellar cooling *sky cooling*	天空冷却	rayonnement sur l'espace *refroidissement par rayonnement terrestre*	1.3.2-**16**
mean radiant temperature	平均辐射温度	température radiante moyenne	1.3.2-**17**
monochromatic emissivity ◐ *spectral emissivity*	光谱发射率 单色辐射率 光谱辐射率	émissivité spectrale *émissivité monochromatique* ◐	1.3.2-**18**
radiance *radiant intensity*	辐射强度	luminance énergétique	1.3.2-**19**
radiant exitance *emissive power* *emittance*	辐射强度	émittance *exitance*	1.3.2-**20**
radiant heat	辐射热	chaleur radiante *chaleur rayonnante*	1.3.2-**21**
radiant intensity *radiance*	辐射强度	luminance énergétique	1.3.2-**22**
radiation	辐射	rayonnement	1.3.2-**23**
radiation heat transfer coefficient	辐射传热系数	coefficient d'échange thermique radiatif	1.3.2-**24**
radiation shield	辐射屏	écran antirayonnement	1.3.2-**25**
radiometer	辐射计	radiomètre	1.3.2-**26**
reflectance *reflectivity* *reflection factor* ◐	反射率	facteur de réflexion *réflectance* *réflectivité*	1.3.2-**27**
reflection factor ◐ *reflectance* *reflectivity*	反射率	facteur de réflexion *réflectance* *réflectivité*	1.3.2-**28**
reflectivity *reflectance* *reflection factor* ◐	反射率	facteur de réflexion *réflectance* *réflectivité*	1.3.2-**29**
shape factor *angle factor* *view factor* *configuration factor* ◐	角系数 形状系数 视角因数	facteur d'angle *facteur de forme*	1.3.2-**30**

	ENGLISH	汉语	FRANÇAIS
1.3.2-**31**	sky cooling *interstellar cooling*	天空冷却	rayonnement sur l'espace *refroidissement par rayonnement terrestre*
1.3.2-**32**	solar constant	太阳常数	constante solaire
1.3.2-**33**	spectral	光谱的 单色的	spectral
1.3.2-**34**	spectral emissivity *monochromatic emissivity* ◐	光谱发射率 单色辐射率 光谱辐射率	émissivité spectrale *émissivité monochromatique* ◐
1.3.2-**35**	spectral emittance *spectral exitance*	光谱发射度	exitance spectrale *émittance spectrale* ◐
1.3.2-**36**	spectral exitance *spectral emittance*	光谱发射度	exitance spectrale *émittance spectrale* ◐
1.3.2-**37**	transmission factor *transmissivity* *transmittance*	透射率	transmittance *transmittivité* *facteur de transmission*
1.3.2-**38**	transmissivity *transmission factor* *transmittance*	透射率	transmittance *transmittivité* *facteur de transmission*
1.3.2-**39**	transmittance *transmission factor* *transmissivity*	透射率	transmittance *transmittivité* *facteur de transmission*
1.3.2-**40**	view factor *angle factor* *shape factor* *configuration factor* ◐	角系数 形状系数 视角因数	facteur d'angle *facteur de forme*

	SECTION 1.3.3 *Thermal conduction*	节 **1.3.3** 导热	**SOUS-CHAPITRE 1.3.3** *Conduction thermique*
1.3.3-**1**	heat conduction *thermal conduction* ◐	热传导	conduction thermique
1.3.3-**2**	thermal conduction ◐ *heat conduction*	热传导	conduction thermique
1.3.3-**3**	thermal conductivity	导热系数	conductivité thermique *coefficient de conductibilité thermique* ○ *coefficient de conduction thermique* ○
1.3.3-**4**	thermal diffusivity	热扩散系数 热扩散率	diffusivité thermique
1.3.3-**5**	thermal resistivity	热阻率	résistivité thermique

	SECTION 1.3.4 *Convection*	节 **1.3.4** 对流换热	**SOUS-CHAPITRE 1.3.4** *Convection*
1.3.4-**1**	convection (of heat)	(热的)对流	convection (de chaleur) *transfert convectif*
1.3.4-**2**	convector (fluid)	对流(流体)	fluide convecteur
1.3.4-**3**	film coefficient of heat transfer ◐ *heat transfer coefficient* *surface coefficient of heat transfer* ◐ *surface film conductance* ◐	对流换热表面传热系数 表面传热系数	coefficient de transfert de chaleur *coefficient d'échange (thermique) superficie* *coefficient de transmission (thermique) de surface*

ENGLISH	汉语	FRANÇAIS	
forced convection	强制对流	convection forcée	1.3.4-**4**
free convection *natural convection*	自然对流	convection naturelle *convection libre*	1.3.4-**5**
heat transfer coefficient *film coefficient of heat transfer* ◐ *surface coefficient of heat transfer* ◐ *surface film conductance* ◐	对流换热表面传热系数 表面传热系数	coefficient de transfert de chaleur *coefficient d'échange (thermique) superficiel* *coefficient de transmission (thermique) de surface*	1.3.4-**6**
natural convection *free convection*	自然对流	convection naturelle *convection libre*	1.3.4-**7**
surface coefficient of heat transfer ◐ *heat transfer coefficient* *film coefficient of heat transfer* ◐ *surface film conductance* ◐	对流换热表面传热系数 表面传热系数	coefficient de transfert de chaleur *coefficient d'échange (thermique) superficiel* *coefficient de transmission thermique de surface*	1.3.4-**8**
surface film conductance ◐ *heat transfer coefficient* *film coefficient of heat transfer* ◐ *surface coefficient of heat transfer* ◐	对流换热表面传热系数 表面传热系数	coefficient de transfert de chaleur *coefficient d'échange (thermique) superficiel* *coefficient de transmission thermique de surface*	1.3.4-**9**

SECTION 1.3.5 *Mass transfer*	节 **1.3.5** 传质	**SOUS-CHAPITRE 1.3.5** *Transfert de masse*	
accommodation coefficient	适应系数	coefficient d'accommodation	1.3.5-**1**
concentration	浓度	concentration	1.3.5-**2**
convective diffusivity *convective mass transfer* *eddy diffusion* *turbulent diffusion*	对流扩散	diffusion convective	1.3.5-**3**
convective mass transfer *convective diffusivity* *eddy diffusion* *turbulent diffusion*	对流扩散	diffusion convective	1.3.5-**4**
diffusion *molecular diffusion*	扩散 分子扩散	diffusion	1.3.5-**5**
diffusion coefficient *mass diffusivity*	扩散系数	coefficient de diffusion (de masse)	1.3.5-**6**
eddy diffusion *convective diffusivity* *convective mass transfer* *turbulent diffusion*	对流扩散	diffusion convective	1.3.5-**7**
mass diffusivity *diffusion coefficient*	扩散系数	coefficient de diffusion (de masse)	1.3.5-**8**
mass transfer *mass transport* ◐	传质	transfert de masse *transport de masse* ◐	1.3.5-**9**
mass transport ◐ *mass transfer*	传质	transfert de masse *transport de masse* ◐	1.3.5-**10**
molecular diffusion *diffusion*	扩散 分子扩散	diffusion	1.3.5-**11**
solute	溶质	soluté	1.3.5-**12**
solvent	溶剂	solvant	1.3.5-**13**
turbulent diffusion *convective diffusivity* *convective mass transfer* *eddy diffusion*	对流扩散	diffusion convective	1.3.5-**14**

	ENGLISH	汉语	FRANÇAIS
	SECTION 1.4 *Related fields* **SECTION 1.4.1** *Acoustics*	节 **1.4** 相关学科 节 **1.4.1** 声学	**SOUS-CHAPITRE 1.4** *Domaines apparentés* **SOUS-CHAPITRE 1.4.1** *Acoustique*
1.4.1-1	acoustic oscillation *sound* *acoustic vibration* ◐	声传播 声音	oscillation acoustique *son* *vibration acoustique* ◐
1.4.1-2	acoustic vibration ◐ *acoustic oscillation* *sound*	声传播 声音	oscillation acoustique *son* *vibration acoustique* ◐
1.4.1-3	ambient noise	环境噪声	bruit ambiant
1.4.1-4	audible sound	可听声	son audible
1.4.1-5	background noise	背景噪声	bruit de fond
1.4.1-6	hertz (Hz)	赫兹 （Hz）	hertz (Hz)
1.4.1-7	infrasound	亚音频	infrason
1.4.1-8	instantaneous sound pressure	瞬时声压	pression acoustique instantanée
1.4.1-9	noise	噪声	bruit
1.4.1-10	octave band	倍频程	bande d'octave
1.4.1-11	one-third-octave band	三分之一倍频带	bande de tiers d'octave
1.4.1-12	pink noise	有色噪声	bruit rose
1.4.1-13	rms sound pressure	均方根声压	pression acoustique efficace
1.4.1-14	sound *acoustic oscillation* *acoustic vibration* ◐	声传播 声音	oscillation acoustique *son* *vibration acoustique* ◐
1.4.1-15	sound power level (L_W)	声功率级 （L_W）	niveau de puissance acoustique (L_W)
1.4.1-16	sound power of a source	源声功率	puissance acoustique d'une source
1.4.1-17	sound pressure	声压	pression acoustique
1.4.1-18	sound pressure level	声压级	niveau de pression acoustique
1.4.1-19	sound spectrum	声谱	spectre acoustique
1.4.1-20	static pressure	静压	pression statique
1.4.1-21	ultrasound	超声(波)	ultrason
1.4.1-22	white noise	白噪声	bruit blanc

	SECTION 1.4.2 *Electricity*	节 **1.4.2** 电学	**SOUS-CHAPITRE 1.4.2** *Electricité*
1.4.2-1	absorbed electrical capacity	消耗电容	puissance électrique absorbée
1.4.2-2	effective power input	有效输入功率	puissance absorbée effective
1.4.2-3	rated frequency	额定频率	fréquence nominale
1.4.2-4	rated voltage	额定电压	tension nominale
1.4.2-5	starting current	起动电流	intensité de démarrage
1.4.2-6	total power input	总输入功率	puissance absorbée totale

章 2. 产冷

◐ 容许（或默许）术语

○ 过时术语

ENGLISH	汉语	FRANÇAIS	
SECTION 2.1 *Heat balance*	节 **2.1** 热平衡	**SOUS-CHAPITRE 2.1** *Bilan thermique*	
cooling load ◐ *heat load* *refrigeration duty* ◐ *refrigeration load* ◐ *refrigeration requirement* ◐	热负荷	charge thermique *besoin de froid* ◐	2.1-**1**
dry tons (USA) ◐ *sensible heat load*	显热负荷	charge thermique "sensible" *charge thermique due à la chaleur sensible* ◐	2.1-**2**
(estimated) design load	(估算)设计负荷	charge frigorifique prévisionnelle	2.1-**3**
heat balance	热平衡	bilan frigorifique *bilan thermique* ◐	2.1-**4**
heat load *cooling load* ◐ *refrigeration duty* ◐ *refrigeration load* ◐ *refrigeration requirement* ◐	热负荷	charge thermique *besoin de froid* ◐	2.1-**5**
latent heat load *moisture tons (USA)* ○ *wet tons* ○	潜热负荷	charge thermique "latente" *charge thermique due à la chaleur latente* ◐	2.1-**6**
load factor	负荷系数	pourcentage de charge *facteur de charge*	2.1-**7**
moisture tons (USA) ○ *latent heat load* *wet tons* ○	潜热负荷	charge thermique "latente" *charge thermique due à la chaleur latente* ◐	2.1-**8**
product load	产品负荷	charge thermique due au produit *besoin de froid pour le produit* ◐	2.1-**9**
refrigeration duty ◐ *heat load* *cooling load* ◐ *refrigeration load* ◐ *refrigeration requirement* ◐	热负荷	charge thermique *besoin de froid* ◐	2.1-**10**
refrigeration load ◐ *heat load* *cooling load* ◐ *refrigeration duty* ◐ *refrigeration requirement* ◐	热负荷	charge thermique *besoin de froid* ◐	2.1-**11**
refrigeration requirement ◐ *heat load* *cooling load* ◐ *refrigeration duty* ◐ *refrigeration load* ◐	热负荷	charge thermique *besoin de froid* ◐	2.1-**12**
sensible heat load *dry tons (USA)* ◐	显热负荷	charge thermique "sensible" *charge thermique due à la chaleur sensible* ◐	2.1-**13**
service load	操作负荷	charges thermiques d'exploitation *charges thermiques d'utilisation* ◐	2.1-**14**
wet tons ○ *latent heat load* *moisture tons (USA)* ○	潜热负荷	charge thermique "latente" *charge thermique due à la chaleur latente* ◐	2.1-**15**

	ENGLISH	汉语	FRANÇAIS
	SECTION 2.2 *Refrigeration capacity and calculation data*	节 **2.2** 制冷量与计算数据	**SOUS-CHAPITRE 2.2** *Puissance frigorifique et éléments de calcul*
2.2-**1**	adiabatic efficiency ◐ *indicated efficiency*	指示效率	rendement indiqué
2.2-**2**	brake horsepower ◐ *power input rating* *shaft horsepower ◐* *shaft power ◐*	轴功率	puissance (mécanique) effective *puissance sur l'arbre ◐*
2.2-**3**	coefficient of performance (COP) *performance energy ratio ◐*	性能系数	coefficient de performance (COP)
2.2-**4**	condenser duty ◐ *condenser heat*	冷凝器热量	chaleur rejetée au condenseur
2.2-**5**	condenser heat *condenser duty ◐*	冷凝器热量	chaleur rejetée au condenseur
2.2-**6**	condensing unit capacity ◐ *overall refrigerating effect*	总产冷量	puissance frigorifique globale
2.2-**7**	degree of superheat	过热度	(degré de) surchauffe
2.2-**8**	effective efficiency ◐ *overall efficiency*	总效率	rendement global *rendement effectif*
2.2-**9**	effective work	有效功	travail effectif
2.2-**10**	efficiency	效率	rendement *efficacité ◐*
2.2-**11**	external conditions	外部条件	régime (thermique) extérieur
2.2-**12**	heat extraction rate	吸热量	transfert de chaleur utile
2.2-**13**	heat of subcooling	过冷热量	chaleur de sous-refroidissement
2.2-**14**	heat recovery capacity	热回收量	puissance thermique de récupération *puissance du récupérateur de chaleur*
2.2-**15**	heat removed	移热量	chaleur enlevée
2.2-**16**	hold-over ◐ *thermal storage*	蓄冷	stockage de froid *accumulation de froid ◐*
2.2-**17**	ice-making capacity	制冰能力	capacité de production de glace
2.2-**18**	indicated efficiency *adiabatic efficiency ◐*	指示效率	rendement indiqué
2.2-**19**	indicated power	指示功率	puissance indiquée
2.2-**20**	indicated work	指示功	travail indiqué
2.2-**21**	internal conditions	内部条件	régime thermique interne
2.2-**22**	internal efficiency	内在效率	rendement interne
2.2-**23**	internal power	内在功率	puissance interne
2.2-**24**	isentropic efficiency	等熵效率	rendement isentropique
2.2-**25**	isentropic power	等熵功率	puissance théorique isentropique
2.2-**26**	isothermal efficiency	等温效率	rendement isothermique *rendement isotherme ◐*
2.2-**27**	isothermal power	等温功率	puissance théorique isotherme *puissance isothermique ◐*
2.2-**28**	mechanical efficiency	机械效率	rendement mécanique
2.2-**29**	net cooling capacity	净冷却能力	puissance frigorifique nette

ENGLISH	汉语	FRANÇAIS	
net refrigerating effect	净产冷量	effet frigorifique net	2.2-**30**
overall efficiency *effective efficiency* ◐	总效率	rendement global *rendement effectif*	2.2-**31**
overall refrigerating effect *condensing unit capacity* ◐	总产冷量	puissance frigorifique globale	2.2-**32**
packaged compressor power input	压缩机组的总输入功率	puissance absorbée d'un groupe motocompresseur	2.2-**33**
performance energy ratio ◐ *coefficient of performance (COP)*	性能系数	coefficient de performance (COP)	2.2-**34**
power input	输入功率	puissance consommée *puissance sur l'arbre* ◐	2.2-**35**
power input rating *brake horsepower* ◐ *shaft horsepower* ◐ *shaft power* ◐	轴功率	puissance (mécanique) effective *puissance sur l'arbre* ◐	2.2-**36**
rated conditions	额定工况	conditions nominales	2.2-**37**
rating under working conditions	实际工况的产冷量	puissance frigorifique effective	2.2-**38**
reference conditions	参考工况	conditions de référence	2.2-**39**
refrigerating capacity	制冷功率（量） 产冷量	puissance frigorifique	2.2-**40**
refrigerating effect per brake horse-power ◐	单位轴功率的产冷量	puissance frigorifique spécifique ◐	2.2-**41**
refrigerating effect per unit of swept volume	单位容积产冷量	production frigorifique volumétrique *production volumétrique effective*	2.2-**42**
(refrigeration) output	(冷)输出量	production (frigorifique)	2.2-**43**
shaft horsepower ◐ *power input rating* *brake horsepower* ◐ *shaft power* ◐	轴功率	puissance (mécanique) effective *puissance sur l'arbre* ◐	2.2-**44**
shaft power ◐ *power input rating* *brake horsepower* ◐ *shaft horsepower* ◐	轴功率	puissance (mécanique) effective *puissance sur l'arbre* ◐	2.2-**45**
specific vaporization enthalpy	比蒸发焓	enthalpie massique d'évaporation	2.2-**46**
standard conditions	标准工况	conditions nominales *conditions de référence*	2.2-**47**
standard rating *total refrigeration* ◐	标准产冷量	puissance (frigorifique) nominale	2.2-**48**
superheat	过热量	chaleur de surchauffe	2.2-**49**
temperature of saturated vapour at the discharge of the compressor	压缩机排气温度	température de vapeur saturée au refoulement du compresseur *température de saturation vapeur à la pression de refoulement*	2.2-**50**
theoretical efficiency	理论效率	rendement théorique	2.2-**51**
thermal storage *hold-over* ◐	蓄冷	stockage de froid *accumulation de froid* ◐	2.2-**52**
total refrigeration ◐ *standard rating*	标准产冷量	puissance (frigorifique) nominale	2.2-**53**
useful refrigerating effect	有效冷量	puissance frigorifique utile	2.2-**54**
volumetric efficiency	容积效率	rendement volumétrique	2.2-**55**

	ENGLISH	汉语	FRANÇAIS
	SECTION 2.3 *Refrigerating systems* **SECTION 2.3.1** *Refrigerating systems: general background*	节 **2.3** 制冷系统 节 **2.3.1** 制冷系统：基础知识	**SOUS-CHAPITRE 2.3** *Systèmes frigorifiques* **SOUS-CHAPITRE 2.3.1** *Généralités sur les systèmes frigorifiques*
2.3.1-**1**	control cycle *cycle period*	受控循环	cycle de fonctionnement
2.3.1-**2**	critical charge	临界充灌量	charge minimale (opérationnelle)
2.3.1-**3**	cycle period *control cycle*	受控循环	cycle de fonctionnement
2.3.1-**4**	direct refrigerating system	直接膨胀制冷系统 直接式制冷系统	système de refroidissement direct *refroidissement par détente directe*
2.3.1-**5**	indirect refrigerating system *secondary cooling system*	间接式制冷系统 间接制冷系统 间接冷却系统	système de refroidissement indirect *système secondaire de refroidissement* *système à fluide secondaire* ◐
2.3.1-**6**	limited charge system	有限充注系统	système à charge contrôlée
2.3.1-**7**	mobile system	可移动的系统	système mobile *système embarqué* ◐
2.3.1-**8**	refrigerating machine ◐ *refrigerating system* *refrigeration system* *refrigeration machine* ◐	制冷系统 制冷机械	système frigorifique *machine frigorifique*
2.3.1-**9**	refrigerating system *refrigeration system* *refrigerating machine* ◐ *refrigeration machine* ◐	制冷系统 制冷机械	système frigorifique *machine frigorifique*
2.3.1-**10**	refrigeration machine ◐ *refrigerating system* *refrigeration system* *refrigerating machine* ◐	制冷系统 制冷机械	système frigorifique *machine frigorifique*
2.3.1-**11**	refrigeration system *refrigerating system* *refrigerating machine* ◐ *refrigeration machine* ◐	制冷系统 制冷机械	système frigorifique *machine frigorifique*
2.3.1-**12**	secondary cooling system *indirect refrigerating system*	间接式制冷系统 间接制冷系统 间接冷却系统	système de refroidissement indirect *système secondaire de refroidissement* *système à fluide secondaire*
2.3.1-**13**	unit system ◐	机组系统	groupe frigorifique fabriqué et essayé en usine *système frigorifique monobloc préfabriqué*

	SECTION 2.3.2 *Compression refrigerating systems*	节 **2.3.2** 压缩制冷系统	**SOUS-CHAPITRE 2.3.2** *Systèmes frigorifiques à compression*
2.3.2-**1**	autorefrigerated cascade ○ *integrated cascade* ◐ *mixed refrigerant cascade* ◐	混合制冷剂复叠系统	cascade intégrée *cascade incorporée* ◐
2.3.2-**2**	compound compression *two-stage compression*	双级压缩	compression biétagée *compression à deux étages* ◐
2.3.2-**3**	compression ratio ◐ *pressure ratio*	压缩比	taux de compression
2.3.2-**4**	compression stage	压缩级	étage de compression

ENGLISH	汉语	FRANÇAIS	
compression system *compression-type refrigerating system*	压缩式制冷系统 压缩系统	système frigorifique à compression *système à compression* ◐	2.3.2-**5**
compression-type refrigerating system *compression system*	压缩式制冷系统 压缩系统	système frigorifique à compression *système à compression* ◐	2.3.2-**6**
condensing pressure	冷凝压力	pression de condensation *pression de liquéfaction* ○	2.3.2-**7**
condensing temperature	冷凝温度	température de condensation	2.3.2-**8**
delivery pressure ○ *discharge pressure*	排气压力	pression de refoulement	2.3.2-**9**
delivery temperature ○ *discharge temperature*	排气温度	température de refoulement	2.3.2-**10**
desuperheating	除过热	désurchauffe *refroidissement avant condensation* ◐	2.3.2-**11**
discharge	排气	refoulement	2.3.2-**12**
discharge pressure *delivery pressure* ○	排气压力	pression de refoulement	2.3.2-**13**
discharge temperature *delivery temperature* ○	排气温度	température de refoulement	2.3.2-**14**
dry compression	干压缩	fonctionnement en régime sec *fonctionnement en régime de surchauffe* ◐	2.3.2-**15**
dual compression ○	双效压缩	compression à double aspiration ○	2.3.2-**16**
evaporating pressure	蒸发压力	pression d'évaporation *pression de vaporisation* ○	2.3.2-**17**
evaporating temperature	蒸发温度	température d'évaporation *température de vaporisation* ○	2.3.2-**18**
external coolant	外界冷却介质	fluide réfrigérant externe *fluide de refroidissement externe*	2.3.2-**19**
high-pressure side *high side* ◐	高压侧	côté haute pression	2.3.2-**20**
high-pressure stage	高压级	étage haute pression	2.3.2-**21**
high side ◐ *high-pressure side*	高压侧	côté haute pression	2.3.2-**22**
inlet pressure	进口压力	pression d'aspiration	2.3.2-**23**
inlet temperature	进口温度	température d'aspiration	2.3.2-**24**
integrated cascade ◐ *mixed refrigerant cascade* ◐ *autorefrigerated cascade* ○	混合制冷剂复叠系统	cascade intégrée *cascade incorporée* ◐	2.3.2-**25**
intercooling *interstage cooling* ◐	中间冷却	refroidissement intermédiaire	2.3.2-**26**
intermediate pressure *interstage pressure* ◐	中间压力	pression intermédiaire	2.3.2-**27**
interstage cooling ◐ *intercooling*	中间冷却	refroidissement intermédiaire	2.3.2-**28**
interstage pressure ◐ *intermediate pressure*	中间压力	pression intermédiaire	2.3.2-**29**
low-pressure side *low side* ◐	低压侧	côté basse pression	2.3.2-**30**
low-pressure stage	低压级	étage basse pression	2.3.2-**31**

	ENGLISH	汉语	FRANÇAIS
2.3.2-**32**	low side ◐ *low-pressure side*	低压侧	côté basse pression
2.3.2-**33**	mixed refrigerant cascade ◐ *integrated cascade* ◐ *autorefrigerated cascade* ○	混合制冷剂复叠系统	cascade intégrée *cascade incorporée* ◐
2.3.2-**34**	multistage compression	多级压缩	compression multiétagée *compression à plusieurs étages* ◐
2.3.2-**35**	multistage expansion	多级膨胀	détente fractionnée *détente étagée* ◐
2.3.2-**36**	pressure ratio *compression ratio* ◐	压缩比	taux de compression
2.3.2-**37**	shaft rotational speed	轴转速	fréquence de rotation (de l'arbre) *vitesse de rotation (de l'arbre)* ◐
2.3.2-**38**	single-stage compression	单级压缩	compression monoétagée *compression à un étage* ◐
2.3.2-**39**	stage pressure ratio	级压缩比	taux de compression par étage *rapport de pression par étage* ◐
2.3.2-**40**	suction	吸气	aspiration
2.3.2-**41**	suction pressure	吸气压力	pression d'aspiration
2.3.2-**42**	suction temperature	吸气温度	température d'aspiration
2.3.2-**43**	total pressure ratio	总压缩比	taux de compression total *rapport total de compression* ◐
2.3.2-**44**	two-stage compression *compound compression*	双级压缩	compression biétagée *compression à deux étages* ◐
2.3.2-**45**	volume factor	容积效率	coefficient de volume *coefficient de débit* ◐
2.3.2-**46**	wet compression	湿压缩	fonctionnement en régime humide *marche en régime humide* ◐ *fonctionnement humide* ◐

	SECTION 2.3.3 *Other refrigerating systems* **SECTION 2.3.3.1** *Sorption system*	节 **2.3.3** 其它制冷系统 节 **2.3.3.1** 吸收（附）系统	**SOUS-CHAPITRE 2.3.3** *Autres systèmes frigorifiques* **SOUS-CHAPITRE 2.3.3.1** *Système à sorption*
2.3.3.1-**1**	absorbate	吸收物	absorbat
2.3.3.1-**2**	absorbent	吸收剂	absorbant
2.3.3.1-**3**	absorber	吸收器	absorbeur
2.3.3.1-**4**	absorption	吸收	absorption
2.3.3.1-**5**	absorption (refrigerating) machine	吸收式（制冷）机	machine (frigorifique) à absorption
2.3.3.1-**6**	absorption refrigerating system	吸收式制冷系统	système frigorifique à absorption
2.3.3.1-**7**	absorption refrigerating unit	吸收式制冷机组	groupe frigorifique à absorption
2.3.3.1-**8**	adsorbate	吸附物	adsorbat
2.3.3.1-**9**	adsorbent	吸附剂	adsorbant
2.3.3.1-**10**	adsorption	吸附	adsorption

ENGLISH	汉语	FRANÇAIS	
analyser	分析器	déphlegmateur	2.3.3.1-**11**
boiler ○ *generator* *regenerator* *desorber* ◐	发生器	générateur *bouilleur* ○	2.3.3.1-**12**
desorber ◐ *generator* *regenerator* *boiler* ○	发生器	générateur *bouilleur* ○	2.3.3.1-**13**
desorption	解吸	désorption	2.3.3.1-**14**
diffusion-absorption system	扩散-吸收式系统	système à (absorption-)diffusion	2.3.3.1-**15**
double-effect cycle	双效循环	cycle à double effet	2.3.3.1-**16**
double-lift cycle	双升（温）循环	cycle biétagé	2.3.3.1-**17**
exhaust steam	排汽	vapeur d'échappement	2.3.3.1-**18**
generator *regenerator* *desorber* ◐ *boiler* ○	发生器	générateur *bouilleur* ○	2.3.3.1-**19**
intermittent absorption refrigerating machine	间歇吸收式制冷机	machine frigorifique à sorption à fonctionnement intermittent *machine frigorifique à absorption discontinue*	2.3.3.1-**20**
liquor ○ *solution*	溶液	solution	2.3.3.1-**21**
multi-effect cycle	多效循环	cycle multi-effets	2.3.3.1-**22**
multi-lift cycle	多升（温）循环	cycle multiétagé	2.3.3.1-**23**
multistage cycle	多级循环	cycle multiétagé	2.3.3.1-**24**
poor solution ◐ *weak solution*	稀溶液	solution faible *solution pauvre* *liqueur pauvre* ○	2.3.3.1-**25**
rectifier	精馏器	colonne de rectification *rectificateur* ◐	2.3.3.1-**26**
regenerator *generator* *desorber* ◐ *boiler* ○	发生器	générateur *bouilleur* ○	2.3.3.1-**27**
resorption system	再吸收式系统	système à résorption	2.3.3.1-**28**
rich solution ◐ *strong solution*	浓溶液	solution riche *solution forte* *liqueur riche* ○	2.3.3.1-**29**
single-effect cycle	单效循环	cycle à simple effet	2.3.3.1-**30**
solution *liquor* ○	溶液	solution	2.3.3.1-**31**
solution width	溶液浓度差	écart de concentration *saut de titre* ◐	2.3.3.1-**32**
sorbate	吸着物	sorbat	2.3.3.1-**33**
sorbent	吸着剂	sorbant	2.3.3.1-**34**
sorption	吸收 吸附	sorption	2.3.3.1-**35**
stripping column	汽提塔	colonne d'épuisement	2.3.3.1-**36**

	ENGLISH	汉语	FRANÇAIS
2.3.3.1-**37**	strong solution *rich solution* ◐	浓溶液	solution riche *solution forte* *liqueur riche* ○
2.3.3.1-**38**	weak solution *poor solution* ◐	稀溶液	solution faible *solution pauvre* *liqueur pauvre* ○

	SECTION 2.3.3.2 *Steam-jet system*	节 **2.3.3.2** 蒸汽喷射系统	**SOUS-CHAPITRE 2.3.3.2** *Système à éjection de vapeur*
2.3.3.2-**1**	ejector	喷射器	ejecto-compresseur
2.3.3.2-**2**	ejector-cycle refrigerating system ◐ *steam-jet refrigerating system*	蒸汽喷射制冷系统	système frigorifique à éjection (de vapeur d'eau)
2.3.3.2-**3**	steam-jet refrigerating machine	蒸汽喷射制冷机	machine frigorifique à éjection (de vapeur d'eau)
2.3.3.2-**4**	steam-jet refrigerating system *ejector-cycle refrigerating system* ◐	蒸汽喷射制冷系统	système frigorifique à éjection (de vapeur d'eau)

	SECTION 2.3.3.3 *Thermoelectric cooling*	节 **2.3.3.3** 热电制冷	**SOUS-CHAPITRE 2.3.3.3** *Refroidissement thermoélectrique*
2.3.3.3-**1**	figure of merit ◐	优值系数	coefficient de mérite *facteur de mérite*
2.3.3.3-**2**	Peltier effect	帕尔帖效应	effet Peltier
2.3.3.3-**3**	Seebeck effect	塞贝克效应	effet Seebeck
2.3.3.3-**4**	semiconductor	半导体	semi-conducteur
2.3.3.3-**5**	thermoelectric battery ○ *thermoelectric module*	热电组件	module thermoélectrique
2.3.3.3-**6**	thermoelectric cooling *thermoelectric refrigeration* ◐	热电制冷	refroidissement thermoélectrique
2.3.3.3-**7**	thermoelectric module *thermoelectric battery* ○	热电组件	module thermoélectrique
2.3.3.3-**8**	thermoelectric refrigeration ◐ *thermoelectric cooling*	热电制冷	refroidissement thermoélectrique
2.3.3.3-**9**	thermopile	热电堆	pile thermoélectrique

	SECTION 2.3.3.4 *Other refrigerating systems*	节 **2.3.3.4** 其它制冷系统	**SOUS-CHAPITRE 2.3.3.4** *Autres systèmes frigorifiques*
2.3.3.4-**1**	air-cycle refrigerating machine	空气循环制冷机	machine frigorifique à air
2.3.3.4-**2**	air (refrigeration) cycle	空气(制冷)循环	cycle (frigorifique) à air
2.3.3.4-**3**	heat pipe	热管	caloduc

ENGLISH	汉语	FRANÇAIS	
pulse-tube ○	脉冲管	tube à pulsation ○	2.3.3.4-**4**
Ranque-Hilsch effect	兰克-赫尔胥效应	effet Ranque-Hilsch	2.3.3.4-**5**
vortex tube	涡流管	tube vortex *tube à tourbillon* ◐	2.3.3.4-**6**

SECTION 2.4 *Compressors* **SECTION 2.4.1** *Compression refrigerating units*	**节 2.4** 压缩机 **节 2.4.1** 压缩制冷机组	**SECTION 2.4** *Compresseurs* **SECTION 2.4.1** *Groupes frigorifiques à compression*	
accessible hermetic motor compressor ○ *semi-hermetic compressor*	半封闭压缩机 半封闭式压缩机	(moto)compresseur hermétique accessible *(moto)compresseur semi-hermétique* ◐	2.4.1-**1**
air-cooled unit	空气冷却式机组	groupe de condensation à air	2.4.1-**2**
commercial condensing unit ◐	商用压缩冷凝机组	groupe frigorifique commercial	2.4.1-**3**
compressor unit ◐	压缩机组	motocompresseur *groupe motocompresseur* ◐	2.4.1-**4**
condensing unit	压缩冷凝机组	unité de condensation *groupe compresseur-condenseur* ◐	2.4.1-**5**
hermetically sealed condensing unit ○	全封闭压缩冷凝机组	groupe frigorifique hermétique	2.4.1-**6**
industrial condensing unit ○	工业用压缩冷凝机组	groupe de condensation industriel	2.4.1-**7**
liquid-cooled unit	液体冷却式机组	groupe frigorifique à condensation par un liquide	2.4.1-**8**
low-capacity condensing unit	分马力压缩冷凝机组	groupe frigorifique de faible puissance ○	2.4.1-**9**
open-type compressor	开启式压缩机	compresseur ouvert	2.4.1-**10**
plug-in unit *plug unit*	跨壁式机组	groupe frigorifique de paroi *groupe frigorifique à insérer* ◐	2.4.1-**11**
plug unit *plug-in unit*	跨壁式机组	groupe frigorifique de paroi *groupe frigorifique à insérer* ◐	2.4.1-**12**
refrigerating unit	制冷机组	groupe frigorifique	2.4.1-**13**
semi-hermetic compressor *accessible hermetic motor compressor* ○	半封闭压缩机 半封闭式压缩机	(moto)compresseur hermétique accessible *(moto)compresseur semi-hermétique* ◐	2.4.1-**14**

SECTION 2.4.2 *Positive-displacement compressors*	**节 2.4.2** 容积式压缩机	**SOUS-CHAPITRE 2.4.2** *Compresseurs volumétriques*	
booster compressor *low-stage compressor* ◐	升压压缩机	compresseur de suralimentation *compresseur basse pression* ◐	2.4.2-**1**
closed crankcase compressor ○ *enclosed compressor* ○	闭式曲轴箱压缩机	compresseur (à carter) fermé	2.4.2-**2**
compound compressor ◐	双级压缩机	compresseur biétagé *compresseur compound* ◐	2.4.2-**3**
compressor	压缩机	compresseur	2.4.2-**4**
diaphragm compressor	隔膜压缩机	compresseur à membrane	2.4.2-**5**

	ENGLISH	汉语	FRANÇAIS
2.4.2-6	differential piston compressor ◐ *stepped piston compound compressor* ○	级差式活塞压缩机	compresseur à pistons différentiels *compresseur à pistons étagés* ◐
2.4.2-7	double-acting compressor	双作用压缩机	compresseur à double effet
2.4.2-8	dry piston compressor	无油活塞压缩机	compresseur sec *compresseur à piston sec* ◐
2.4.2-9	dual compressor ◐ *tandem compressor* ◐	串轴压缩机 串轴式压缩机	compresseurs jumelés *compresseurs en tandem* ◐
2.4.2-10	dual-effect compressor	双吸气压缩机	compresseur avec orifice de suralimentation
2.4.2-11	enclosed compressor ○ *closed crankcase compressor* ○	闭式曲轴箱压缩机	compresseur (à carter) fermé
2.4.2-12	liquid-injected rotary compressor	喷液回转压缩机	compresseur rotatif à injection de fluide
2.4.2-13	low-stage compressor ◐ *booster compressor*	升压压缩机	compresseur de suralimentation *compresseur basse pression* ◐
2.4.2-14	monoscrew compressor *single-screw compressor*	单螺杆压缩机	compresseur à vis monomoteur
2.4.2-15	multistage compressor	多级压缩机	compresseur multiétagé
2.4.2-16	multivane rotary compressor	多滑片回转式压缩机	compresseur rotatif à palettes multiples *compresseur rotatif multicellulaire* ○
2.4.2-17	oil-free compressor	无油压缩机	compresseur non lubrifié
2.4.2-18	open crankcase compressor ◐	开启式压缩机	compresseur (à carter) ouvert ◐
2.4.2-19	open-type compressor	开启式压缩机	compresseur ouvert
2.4.2-20	positive-displacement compressor ◐	容积压缩机	compresseur volumétrique
2.4.2-21	reciprocating compressor	往复压缩机	compresseur alternatif *compresseur à piston*
2.4.2-22	refrigerant compressor	制冷压缩机	compresseur frigorifique
2.4.2-23	return-flow compressor ○	逆流压缩机	compresseur à flux alternatif
2.4.2-24	rolling piston compressor	滚动活塞压缩机	compresseur à piston tournant *compresseur à piston rotatif* ◐
2.4.2-25	rotary compressor	回转压缩机	compresseur rotatif
2.4.2-26	screw compressor	螺杆压缩机	compresseur à vis *compresseur hélicoïde* ◐
2.4.2-27	scroll compressor	涡旋压缩机	compresseur à spirale *compresseur spiro-orbital* ◐
2.4.2-28	single acting compressor	单作用压缩机	compresseur à simple effet
2.4.2-29	single-screw compressor *monoscrew compressor*	单螺杆压缩机	compresseur à vis monomoteur
2.4.2-30	single-vane rotary compressor ◐	单滑片回转式压缩机	compresseur rotatif à palette unique *compresseur rotatif monocellulaire* ○
2.4.2-31	sliding-vane compressor ◐	滑片式压缩机	compresseur à palettes
2.4.2-32	standard discharge point	标准排出点	orifice de refoulement *refoulement*
2.4.2-33	standard inlet point	标准吸入点	orifice d'aspiration *aspiration*
2.4.2-34	stepped piston compound compressor ○ *differential piston compressor* ◐	级差式活塞压缩机	compresseur à pistons différentiels *compresseur à pistons étagés* ◐
2.4.2-35	swash plate compressor	斜盘式压缩机	compresseur volumétrique à barillet

ENGLISH	汉语	FRANÇAIS	
tandem compressor ◐ *dual compressor* ◐	串轴压缩机 串轴式压缩机	compresseurs jumelés *compresseurs en tandem* ◐	2.4.2-**36**
twin cylinder compressor	双缸压缩机	compresseur à deux cylindres	2.4.2-**37**
uniflow compressor ○	顺流式压缩机	compresseur à flux continu *compresseur à équicourant* ◐	2.4.2-**38**
variable displacement wobble plate compressor	变排量斜盘式压缩机	compresseur volumétrique variable à plateau oscillant	2.4.2-**39**

SECTION 2.4.3 *Components of positive-displacement compressors*	节 **2.4.3** 容积式压缩部件	**SOUS-CHAPITRE 2.4.3** *Eléments des compresseurs volumétriques*	
actual displacement	实际排气量	débit volume aspiré (m^3/h) *débit volume déplacé réel* ◐	2.4.3-**1**
ball bearing	滚珠轴承	palier à billes *roulement à billes* ◐	2.4.3-**2**
beam valve *leaf valve*	条状阀	soupape à lamelle(s)	2.4.3-**3**
bearing	轴承	palier *coussinet*	2.4.3-**4**
bearing housing	轴承箱	boîtier de palier *cage de palier* ◐	2.4.3-**5**
bellows seal	波纹管密封	garniture d'étanchéité à soufflet	2.4.3-**6**
big end	（连杆） 大头	tête de bielle	2.4.3-**7**
cantilever valve ◐ *flapper valve*	舌状阀	soupape à lamelles multiples *soupape à languette* ◐ *clapet fléchissant* ◐ *clapet battant* ◐	2.4.3-**8**
clearance pocket ○	附加余隙	espace mort additionnel *poche d'espace mort* ○	2.4.3-**9**
clearance volume	余隙容积	espace mort	2.4.3-**10**
clearance volume ratio	相对余隙容积	pourcentage d'espace mort *taux d'espace mort* ◐	2.4.3-**11**
compression stroke	压缩行程	course de compression *compression (phase de)*	2.4.3-**12**
(compressor) valve	（压缩机的） 气阀	soupape *clapet* ◐	2.4.3-**13**
connecting rod	连杆	bielle	2.4.3-**14**
coupling (of shafts)	联轴节	accouplement direct	2.4.3-**15**
crankcase	曲轴箱	carter	2.4.3-**16**
crankcase seal *(shaft) seal*	轴封	garniture d'étanchéité	2.4.3-**17**
crankpin	曲柄销	maneton	2.4.3-**18**
crankshaft	曲轴	vilebrequin	2.4.3-**19**
cylinder	气缸	cylindre	2.4.3-**20**
cylinder block	气缸体	bloc cylindre	2.4.3-**21**

	ENGLISH	汉语	FRANÇAIS
2.4.3-**22**	(cylinder) bore	缸径	alésage
2.4.3-**23**	cylinder head	气缸头	culasse *tête de cylindre* ◐
2.4.3-**24**	cylinder liner	气缸套	chemise de cylindre amovible
2.4.3-**25**	cylinder wall	气缸壁	paroi (intérieure) du cylindre *surface intérieure du cylindre*
2.4.3-**26**	delivery valve ◐ *discharge valve* *outlet valve* ○	排气阀	soupape de refoulement
2.4.3-**27**	diaphragm valve ◐ *disc valve*	园盘阀	clapet à disque
2.4.3-**28**	disc valve *diaphragm valve* ◐	园盘阀	clapet à disque
2.4.3-**29**	discharge line valve ◐ *discharge stop valve* ◐	排气截止阀	robinet de refoulement *robinet d'arrêt au refoulement* ◐
2.4.3-**30**	discharge stop valve ◐ *discharge line valve* ◐	排气截止阀	robinet de refoulement *robinet d'arrêt au refoulement* ◐
2.4.3-**31**	discharge stroke ◐	排气行程	course de refoulement
2.4.3-**32**	discharge valve *delivery valve* ◐ *outlet valve* ○	排气阀	soupape de refoulement
2.4.3-**33**	displacement	排气量	volume balayé *débit volumétrique* ◐
2.4.3-**34**	eccentric strap	偏心环	collier d'excentrique
2.4.3-**35**	economizer connexion *side port* ◐	经济器接口	orifice de suralimentation *orifice économiseur* ◐
2.4.3-**36**	expansion stroke	膨胀行程	course de détente
2.4.3-**37**	flapper valve *cantilever valve* ◐	舌状阀	soupape à lamelles multiples *soupape à languette* ◐ *clapet fléchissant* ◐ *clapet battant* ◐
2.4.3-**38**	flexing valve ◐ *reed valve* ◐	簧片阀	soupape à lamelles multiples *soupape flexible* ◐ *soupape à ruban* ◐
2.4.3-**39**	footstep bearing	臼形轴承	palier de butée vertical *crapaudine* ◐
2.4.3-**40**	friction ring ◐ *rubbing ring*	摩擦环	bague d'étanchéité *anneau d'étanchéité* ◐
2.4.3-**41**	gland packing ◐ *shaft packing* ◐	轴封填料	tresse graphitée d'étanchéité *garniture d'étanchéité*
2.4.3-**42**	gudgeon pin ◐ *piston pin* *wrist pin* ◐	活塞销	axe de piston *axe de pied de bielle* ◐
2.4.3-**43**	(guided) vane	（导向）叶片	palette libre
2.4.3-**44**	hydraulic thrust	平衡活塞	piston d'équilibrage
2.4.3-**45**	journal	轴颈	tourillon
2.4.3-**46**	labyrinth seal	迷宫密封	labyrinthe
2.4.3-**47**	leaf valve *beam valve*	条状阀	soupape à lamelle(s)
2.4.3-**48**	little end *small end* ◐	（连杆）小头	pied de bielle

ENGLISH	汉语	FRANÇAIS	
(mechanical) clearance	(机械的）间隙	jeu (mécanique)	2.4.3-**49**
mechanical seal	机械密封	garniture mécanique	2.4.3-**50**
mushroom valve	菌状阀	soupape tronconique *soupape champignon* ◐	2.4.3-**51**
needle bearing	滚针轴承	palier à aiguilles	2.4.3-**52**
outlet valve ○ *discharge valve* *delivery valve* ◐	排气阀	soupape de refoulement	2.4.3-**53**
partial duty port	部分负荷旁通口	lumière de réduction de puissance	2.4.3-**54**
pedestal bearing	外轴承	palier support auxiliaire	2.4.3-**55**
piston	活塞	piston	2.4.3-**56**
piston displacement ◐ *swept volume* ◐	活塞排量	cylindrée	2.4.3-**57**
piston pin *gudgeon pin* ◐ *wrist pin* ◐	活塞销	axe de piston *axe de pied de bielle* ◐	2.4.3-**58**
piston ring	活塞环	segment d'étanchéité	2.4.3-**59**
piston stroke	活塞行程	course	2.4.3-**60**
poppet valve	提升阀	soupape à queue	2.4.3-**61**
reed valve ◐ *flexing valve* ◐	簧片阀	soupape à lamelles multiples *soupape flexible* ◐ *soupape à ruban* ◐	2.4.3-**62**
ring plate valve ◐ *ring valve*	环片阀	soupape annulaire *soupape concentrique*	2.4.3-**63**
ring valve *ring plate valve* ◐	环片阀	soupape annulaire *soupape concentrique*	2.4.3-**64**
roller bearing	滚柱轴承	palier à rouleaux	2.4.3-**65**
rotary seal	转动密封	garniture d'étanchéité mécanique	2.4.3-**66**
rubbing ring *friction ring* ◐	摩擦环	bague d'étanchéité *anneau d'étanchéité* ◐	2.4.3-**67**
safety (cylinder) head	安全（气缸）盖	fond mobile (sécurité anti-coup de liquide) *tête de cylindre à sécurité* ◐	2.4.3-**68**
safety (valve) head ◐	安全（阀）盖	soupape de refoulement de sécurité ◐	2.4.3-**69**
scraper ring	刮油环	segment racleur	2.4.3-**70**
shaft packing ◐ *gland packing* ◐	轴封填料	tresse graphitée d'étanchéité *garniture d'étanchéité*	2.4.3-**71**
(shaft) seal *crankcase seal*	轴封	garniture d'étanchéité	2.4.3-**72**
side port ◐ *economizer connexion*	经济器接口	orifice de suralimentation *orifice économiseur* ◐	2.4.3-**73**
sleeve bearing	滑动轴承	palier lisse	2.4.3-**74**
slide valve	滑阀	tiroir de variation de puissance *tiroir de variation de volume*	2.4.3-**75**
small end ◐ *little end*	（连杆）小头	pied de bielle	2.4.3-**76**
stuffing box	填料盒	presse-étoupe (garniture de)	2.4.3-**77**
suction inlet	吸气口	orifice d'aspiration	2.4.3-**78**

	ENGLISH	汉语	FRANÇAIS
2.4.3-**79**	suction line valve *suction stop valve* ◐	吸气截止阀	robinet d'aspiration
2.4.3-**80**	suction stop valve ◐ *suction line valve*	吸气截止阀	robinet d'aspiration
2.4.3-**81**	suction stroke	吸气行程	course d'aspiration
2.4.3-**82**	suction valve	吸气阀	clapet d'aspiration *soupape d'aspiration* ◐
2.4.3-**83**	swept volume ◐ *piston displacement* ◐	活塞排量	cylindrée
2.4.3-**84**	theoretical displacement (1)	理论排量	volume engendré
2.4.3-**85**	theoretical displacement (2)	理论排量	volume balayé *volume engendré*
2.4.3-**86**	thrust bearing	止推轴承	palier de butée
2.4.3-**87**	thrust collar	止推盘	collet de butée
2.4.3-**88**	tulip valve	漏斗形阀	soupape tulipe
2.4.3-**89**	valve cage	阀盖	boîtier à soupape *boîte à soupape* *cage de soupape*
2.4.3-**90**	valve cover	阀罩	couvercle de soupape *chapeau de soupape* ○
2.4.3-**91**	valve guard	升程限制器	contre clapet *butée de soupape* ◐
2.4.3-**92**	valve lift	阀升程	levée *levée de clapet* ◐
2.4.3-**93**	valve plate	阀板	plaque porte-clapet *plaque porte-soupape* ◐
2.4.3-**94**	valve port	阀孔	lumière de passage
2.4.3-**95**	valve seat	阀座	siège de soupape
2.4.3-**96**	valve stem	阀杆	tige de soupape
2.4.3-**97**	Vi *volume index*	内容积比	volume index *Vi*
2.4.3-**98**	volume index *Vi*	内容积比	volume index *Vi*
2.4.3-**99**	water (cooled) jacket	水套	chemise d'eau *culasse refroidie par eau* ◐
2.4.3-**100**	wrist pin ◐ *piston pin* *gudgeon pin* ◐	活塞销	axe de piston *axe de pied de bielle* ◐

	SECTION 2.4.4 *Turbocompressors*	节 **2.4.4** 透平压缩机	**SOUS-CHAPITRE 2.4.4** *Turbocompresseurs*
2.4.4-**1**	axial flow compressor	轴流式压缩机	turbocompresseur axial
2.4.4-**2**	blade *vane* ◐	叶片	ailette *aube* *pale* ◐
2.4.4-**3**	blade flutter	叶片颤振	vibration d'ailette *vibration de pale* ◐

ENGLISH	汉语	FRANÇAIS	
blade passing frequency	叶片通过频率	fréquence due aux sillages d'ailettes	2.4.4-**4**
casing (of a turbocompressor)	（透平压缩机的）机壳	corps de compresseur	2.4.4-**5**
centrifugal compressor	离心式压缩机	compresseur centrifuge	2.4.4-**6**
diffuser	扩压器	diffuseur	2.4.4-**7**
drag coefficient	阻力系数	trainée	2.4.4-**8**
eye ◐ *inlet*	进气口	ouïe (d'entrée)	2.4.4-**9**
flexible-shaft centrifugal compressor	挠性轴离心式压缩机	compresseur centrifuge à arbre flexible *compresseur à arbre souple* ◐	2.4.4-**10**
flow coefficient for a dynamic compressor	动力压缩机的流动系数	invariant de débit d'un étage de turbo-compresseur	2.4.4-**11**
impeller *rotor* *wheel* ○	叶轮 转子	roue *rotor* *mobile* ◐	2.4.4-**12**
impeller reaction	叶轮反应度	degré de réaction (dans la roue) *taux de réaction (dans la roue)* ◐	2.4.4-**13**
impeller running frequency	叶轮转动频率	fréquence sonore	2.4.4-**14**
inlet *eye* ◐	进气口	ouïe (d'entrée)	2.4.4-**15**
non-positive displacement compressor	非容积式压缩机	compresseur cinétique	2.4.4-**16**
open impeller *unshrouded impeller* ◐	半开式叶轮	roue ouverte	2.4.4-**17**
overall pressure coefficient for a dynamic compressor	速度型压缩机的总压力系数	invariant de pression	2.4.4-**18**
pre-rotary vane ◐ *pre-rotation vane*	导流叶片	aube de prérotation *ventelle* ◐	2.4.4-**19**
pre-rotation vane *pre-rotary vane* ◐	导流叶片	aube de prérotation *ventelle* ◐	2.4.4-**20**
pre-rotary vane assembly	导流叶片组件	aubage (directeur) de prérotation	2.4.4-**21**
rotor *impeller* *wheel* ○	叶轮 转子	roue *rotor* *mobile* ◐	2.4.4-**22**
scroll *volute* ◐	蜗壳	volute	2.4.4-**23**
shrouded impeller	闭式叶轮	roue fermée	2.4.4-**24**
slip	滑动	décollement	2.4.4-**25**
stall (of a turbocompressor or centrifugal compressor) *stalling* ◐	（透平压缩机的）脱流	blocage	2.4.4-**26**
stalling ◐ *stall (of a turbocompressor or centrifugal compressor)*	（透平压缩机的）脱流	blocage	2.4.4-**27**
stiff-shaft centrifugal compressor	刚性轴离心式压缩机	compresseur centrifuge à arbre rigide	2.4.4-**28**
subsonic compressor	亚音速压缩机	compresseur subsonique	2.4.4-**29**
supersonic compressor	超音速压缩机	compresseur supersonique	2.4.4-**30**
surging (of a turbocompressor or centrifugal compressor)	（透平机的）喘振	pompage (d'une turbomachine)	2.4.4-**31**

	ENGLISH	汉语	FRANÇAIS
2.4.4-**32**	swirl	涡流	tourbillon
2.4.4-**33**	total to static efficiency ◐	静压效率	rendement isentropique
2.4.4-**34**	total to total efficiency ◐	全压效率	rendement polytropique
2.4.4-**35**	turbocompressor	透平压缩机	turbocompresseur *turbomachine*
2.4.4-**36**	unshrouded impeller ◐ *open impeller*	半开式叶轮	roue ouverte
2.4.4-**37**	vane ◐ *blade*	叶片	ailette *aube* *pale* ◐
2.4.4-**38**	volute ◐ *scroll*	蜗壳	volute
2.4.4-**39**	wheel ○ *impeller* *rotor*	叶轮 转子	roue *rotor* *mobile* ◐

	SECTION 2.5 *Heat exchangers* **SECTION 2.5.1** *Heat exchangers: general background*	节 **2.5** 换热器 节 **2.5.1** 换热器：基础知识	**SOUS-CHAPITRE 2.5** *Echangeurs thermiques* **SOUS-CHAPITRE 2.5.1** *Généralités sur les échangeurs thermiques*
2.5.1-**1**	baffle ◐ *baffle plate*	折流板	chicane
2.5.1-**2**	baffle plate *baffle* ◐	折流板	chicane
2.5.1-**3**	baudelot cooler ○ *irrigation cooler* ◐ *surface liquid cooler* ◐	表面液体冷却器	refroidisseur à ruissellement
2.5.1-**4**	bottle-type liquid cooler	瓶式液体冷却器	fontaine réfrigérée
2.5.1-**5**	brazed plate heat exchanger	钎焊板式换热器	échangeur à plaques *échangeur à plaques brasées ou soudées*
2.5.1-**6**	brushed surface	刮刷表面	surface brossée
2.5.1-**7**	bundle of tubes	管束	faisceau tubulaire *faisceau de tubes* ◐
2.5.1-**8**	coil depth *core depth*	盘管深度	profondeur ailetée (d'une batterie) *profondeur*
2.5.1-**9**	coil face area *core face area*	盘管迎风面积	section frontale
2.5.1-**10**	coil length *core length*	盘管长度	longueur ailetée (d'une batterie)
2.5.1-**11**	coil width *core width*	盘管宽度	largeur ailetée (d'une batterie)
2.5.1-**12**	compact heat exchanger	紧凑式换热器	échangeur (thermique) compact
2.5.1-**13**	cooling surface	冷却表面	surface froide
2.5.1-**14**	core depth *coil depth*	盘管深度	profondeur ailetée (d'une batterie) *profondeur*

ENGLISH	汉语	FRANÇAIS	
core face area *coil face area*	盘管迎风面积	section frontale	2.5.1-**15**
core length *coil length*	盘管长度	longueur ailetée (d'une batterie)	2.5.1-**16**
core width *coil width*	盘管宽度	largeur ailetée (d'une batterie)	2.5.1-**17**
counterflow arrangement	逆流排列	circulation à contre-courant	2.5.1-**18**
counterflow exchange	逆流交换	échange à contre-courant	2.5.1-**19**
counterflow heat exchanger	逆流换热器	échangeur (thermique) à contrecourant	2.5.1-**20**
cross-flow exchange	叉流交换	échange à courants croisés	2.5.1-**21**
cross-flow heat exchanger	叉流换热器	échangeur (thermique) à courants croisés	2.5.1-**22**
desuperheater	降低过热度的换热器	désurchauffeur	2.5.1-**23**
direct-contact heat exchanger	直接接触式换热器	échangeur (thermique) à contact direct *échangeur par mélange*	2.5.1-**24**
direct surface ○ *primary surface* ◐	基表面	surface primaire *surface directe*	2.5.1-**25**
drum cooler	鼓式冷却器	refroidisseur à tambour rotatif	2.5.1-**26**
dry cooler	干式冷却器	refroidisseur de fluide atmosphérique sec *aérorefroidisseur sec* ◐	2.5.1-**27**
enclosed liquid cooler	封闭液体冷却器	refroidisseur de liquide à évaporation extratubulaire	2.5.1-**28**
enhanced surface	换热强化表面	tubes à surface améliorée	2.5.1-**29**
extended surface	扩展表面	surface "augmentée" *surface "étendue"*	2.5.1-**30**
face area	迎风面积	section frontale	2.5.1-**31**
fin	翅片 肋片	ailette	2.5.1-**32**
fin efficiency	肋效率	rendement d'ailette	2.5.1-**33**
fin pitch	（每英寸）肋片数	nombre d'ailettes par unité de longueur	2.5.1-**34**
finned length	肋管长度	longueur ailetée *longueur utile d'un tube à ailettes*	2.5.1-**35**
finned surface	肋（翅）化表面	surface ailetée *surface des ailettes* ◐	2.5.1-**36**
finned tube	肋（翅）片管 翅片管	tube à ailettes	2.5.1-**37**
flow area *free flow area*	流通面积	section de passage *section claire* ◐	2.5.1-**38**
fouling	污垢	encrassement	2.5.1-**39**
fouling resistance	污垢热阻	résistance thermique d'encrassement	2.5.1-**40**
free flow area *flow area*	流通面积	section de passage *section claire* ◐	2.5.1-**41**
gas cooler	气体冷却器	refroidisseur de gaz	2.5.1-**42**
grid coil ○ *hairpin coil* ◐	蛇形管	échangeur en épingle à cheveux	2.5.1-**43**
hairpin coil ◐ *grid coil* ○	蛇形管	échangeur en épingle à cheveux	2.5.1-**44**

	ENGLISH	汉语	FRANÇAIS
2.5.1-**45**	heat exchange surface	换热面积	surface d'échange (thermique)
2.5.1-**46**	heat exchanger	换热器	échangeur thermique *échangeur de chaleur* ◐
2.5.1-**47**	heat recovery heat exchanger	热回收换热器	échangeur de récupération de chaleur
2.5.1-**48**	heat recovery liquid chilling package	热回收（用）液体冷却机组	groupe refroidisseur de liquide récupérateur de chaleur
2.5.1-**49**	heat transfer surface (air side)	（空气侧）传热表面	surface d'échange de chaleur (côté air)
2.5.1-**50**	heating surface	加热表面	surface de chauffe
2.5.1-**51**	helical fin ◐ *spiral fin* ◐ *strip fin* ◐	螺旋肋片 螺旋肋	ailette spiralée *ailette hélicoïdale* ◐
2.5.1-**52**	ice bank cooler	蓄冰式冷却器	refroidisseur d'eau accumulateur de glace
2.5.1-**53**	in-line bank of tubes	顺排管组	nappe de tubes alignés
2.5.1-**54**	indirect surface ◐ *secondary surface*	二次表面	surface secondaire *surface indirecte*
2.5.1-**55**	inlet temperature	进口温度	température d'entrée
2.5.1-**56**	inner fin	内肋	ailette interne *ailette intérieure* ◐
2.5.1-**57**	insert ◐ *turbulator*	紊流器	turbulateur *insert* ◐
2.5.1-**58**	integral fin	整体肋	ailette extrudée
2.5.1-**59**	internal volume	内容积	volume interne
2.5.1-**60**	irrigation cooler ◐ *surface liquid cooler* ◐ *baudelot cooler* ○	表面液体冷却器	refroidisseur à ruissellement
2.5.1-**61**	liquid cooler	液体冷却器	refroidisseur de liquide
2.5.1-**62**	liquid inlet temperature	液体进口温度	température du liquide à l'entrée
2.5.1-**63**	liquid outlet temperature	液体出口温度	température du liquide à la sortie
2.5.1-**64**	longitudinal fin	纵肋	ailette longitudinale
2.5.1-**65**	method of heat transfer	传热方法	mode d'échange thermique
2.5.1-**66**	microchannel heat exchanger	微通道换热器	échangeur (thermique) à micro canaux
2.5.1-**67**	multipipe heat exchanger ◐ *multitubular heat exchanger*	多管式换热器	échangeur (thermique) multitubulaire
2.5.1-**68**	multichannel heat exchanger	多通道换热器	échangeur (thermique) multicanaux
2.5.1-**69**	multitubular heat exchanger *multipipe heat exchanger* ◐	多管式换热器	échangeur (thermique) multitubulaire
2.5.1-**70**	number of passes	通道数	nombre de passes
2.5.1-**71**	outlet temperature	出口温度	température de sortie
2.5.1-**72**	parallel flow arrangement	顺流排列	circulation à cocourant *circulation à équicourant* ◐
2.5.1-**73**	parallel flow heat exchanger	顺流换热器	échangeur (thermique) à cocourant *échangeur (thermique) à équicourant* ◐
2.5.1-**74**	(pipe) coil	盘管 排管	serpentin
2.5.1-**75**	plain tube *smooth tube* *smooth pipe* ◐	光管	tube lisse

ENGLISH	汉语	FRANÇAIS	
plate fin	套片	ailette multiple	2.5.1-**76**
plate heat exchanger	板式换热器	échangeur (thermique) à plaques	2.5.1-**77**
plate liquid cooler	平板液体冷却器	refroidisseur de liquide à plaques	2.5.1-**78**
primary surface ◐ *direct surface* ○	基表面	surface primaire *surface directe*	2.5.1-**79**
removable end cover *removable head* ○	可拆端盖	fond démontable *boîte d'extrémité démontable* ◐	2.5.1-**80**
removable head ○ *removable end cover*	可拆端盖	fond démontable *boîte d'extrémité démontable* ◐	2.5.1-**81**
rotary heat exchanger	旋转式换热器	échangeur thermique rotatif	2.5.1-**82**
row	排数	rangée	2.5.1-**83**
row of tubes ◐	管排	nappe de tubes *rideau de tubes*	2.5.1-**84**
scraped heat exchanger	刮削式换热器	échangeur thermique à surface raclée	2.5.1-**85**
secondary surface *indirect surface* ◐	二次表面	surface secondaire *surface indirecte*	2.5.1-**86**
shell-and-tube heat exchanger	壳管式换热器	échangeur (thermique) à (tubes et) calandre	2.5.1-**87**
smooth pipe ◐ *plain tube* *smooth tube*	光管	tube lisse	2.5.1-**88**
smooth tube *plain tube* *smooth pipe* ◐	光管	tube lisse	2.5.1-**89**
spiral fin ◐ *helical fin* ◐ *strip fin* ◐	螺旋肋片 螺旋肋	ailette spiralée *ailette hélicoïdale* ◐	2.5.1-**90**
staggered bank of tubes	错排管组	nappes de tubes en quinconce	2.5.1-**91**
strip fin ◐ *helical fin* ◐ *spiral fin* ◐	螺旋肋片 螺旋肋	ailette spiralée *ailette hélicoïdale* ◐	2.5.1-**92**
surface liquid cooler ◐ *irrigation cooler* ◐ *baudelot cooler* ○	表面液体冷却器	refroidisseur à ruissellement	2.5.1-**93**
(thermal) recuperator *(thermal) regenerator*	蓄热器 蓄冷器	récupérateur thermique *régénérateur thermique* ◐	2.5.1-**94**
(thermal) regenerator *(thermal) recuperator*	蓄热器 蓄冷器	récupérateur thermique *régénérateur thermique* ◐	2.5.1-**95**
tube plate ◐ *tube sheet*	管板	plaque tubulaire	2.5.1-**96**
tube sheet *tube plate* ◐	管板	plaque tubulaire	2.5.1-**97**
turbulator *insert* ◐	紊流器	turbulateur *insert* ◐	2.5.1-**98**
water chiller *water cooler* ◐	冷水机组	refroidisseur d'eau	2.5.1-**99**
water cooler ◐ *water chiller*	冷水机组	refroidisseur d'eau	2.5.1-**100**
welded plate heat exchanger	焊接板式换热器	échangeur à cassettes	2.5.1-**101**

	ENGLISH	汉语	FRANÇAIS
	SECTION 2.5.2 *Condensers and condenser accessories*	节 **2.5.2** 冷凝器与冷凝器附件	**SOUS-CHAPITRE 2.5.2** *Condenseurs et accessoires*
2.5.2-**1**	air-cooled condenser	空冷式冷凝器	condenseur à air *aérocondenseur* ◐
2.5.2-**2**	algaecide	防藻剂	algicide
2.5.2-**3**	approach	冷幅高	approche
2.5.2-**4**	atmospheric condenser	淋激式冷凝器	condenseur à ruissellement
2.5.2-**5**	atmospheric cooling tower ○ *natural-draught cooling tower*	自然通风冷却塔	refroidisseur d'eau atmosphérique à tirage nature *tour de refroidissement à tirage naturel*
2.5.2-**6**	back-up valve	升压阀	robinet à contre pression
2.5.2-**7**	barometric condenser	混合式冷凝器	condenseur barométrique
2.5.2-**8**	bleeder-type condenser	溢流式冷凝器	condenseur à soutirage
2.5.2-**9**	blow-down water	喷射水	purge de déconcentration
2.5.2-**10**	circulating water	循环水	eau de circulation
2.5.2-**11**	closed shell-and-tube condenser	卧式（壳管）冷凝器	condenseur à calandre fermé
2.5.2-**12**	cold water	冷却水	eau refroidie
2.5.2-**13**	cold water basin	冷却水槽	bassin d'eau refroidie
2.5.2-**14**	condenser	冷凝器	condenseur
2.5.2-**15**	condenser-evaporator	冷凝器-蒸发器	condenseur évaporateur
2.5.2-**16**	condenser-receiver	冷凝-贮液器	condenseur-réservoir *condenseur accumulateur de liquide* ◐
2.5.2-**17**	condenser subcooling	冷凝器过冷	sous- refroidissement (en sortie de condenseur)
2.5.2-**18**	cooling tower	冷却塔	refroidisseur d'eau atmosphérique *tour de refroidissement (d'eau)*
2.5.2-**19**	desuperheating coil	预冷盘管	échangeur désurchauffeur *élément désurchauffeur* ◐
2.5.2-**20**	dry cooling tower	干式冷却塔	tour de refroidissement sèche *refroidisseur d'eau atmosphérique sec*
2.5.2-**21**	evaporative condenser	蒸发式冷凝器	condenseur évaporatif *condenseur à évaporation d'eau* ◐ *condenseur évaporateur d'eau* ◐
2.5.2-**22**	fill (of a cooling tower) *packing (of a cooling tower)*	冷却塔填料	garnissage (d'un refroidisseur atmosphérique, d'une tour de refroidissement *remplissage (d'un refroidisseur atmosphérique, d'une tour de refroidissement)* *matelas dispersant*
2.5.2-**23**	film cooling tower	水膜式冷却塔	tour de refroidissement à écoulement pelliculaire
2.5.2-**24**	film packing	膜式填料	garnissage à écoulement pelliculaire
2.5.2-**25**	forced-draught condenser	强制通风式冷凝器	condenseur à air à convection forcée
2.5.2-**26**	forced-draught cooling tower	强制通风式冷却塔	refroidisseur d'eau atmosphérique à air forcé *tour de refroidissement à air forcé*
2.5.2-**27**	grid fill ◐ *grid packing*	格栅形填料	garnissage en treillage *garnissage en grille*

ENGLISH	汉语	FRANÇAIS	
grid packing *grid fill* ◐	格栅形填料	garnissage en treillage *garnissage en grille*	2.5.2-**28**
heat rejection capacity	排热量	puissance thermique évacuée	2.5.2-**29**
hot water *warm water*	热水	eau chaude	2.5.2-**30**
hot water basin *warm water basin*	热水槽	bassin d'accumulation d'eau chaude	2.5.2-**31**
induced draught cooling tower	吸风式冷却塔	refroidisseur d'eau atmosphérique à ventilation par aspiration *tour de refroidissement à ventilation par aspiration* *refroidisseur d'eau atmosphérique à air induit* ◐ *tour de refroidissement à air induit* ◐	2.5.2-**32**
liquefier	液化器	liquéfacteur	2.5.2-**33**
liquid-cooled refrigerant condenser	液冷式冷凝器	condenseur refroidi par un liquide	2.5.2-**34**
make-up water	补水	eau d'appoint	2.5.2-**35**
mechanical draught cooling tower	机械通风式冷却塔	refroidisseur d'eau atmosphérique à ventilation forcée *tour de refroidissement à ventilation forcée*	2.5.2-**36**
multishell condenser ◐	组筒式冷凝器	condenseur à calandres multiples *condenseur multitubulaire à plusieurs corps* ◐	2.5.2-**37**
(multitubular) shell-and-tube condenser ◐ *shell-and-tube condenser*	（多管）壳管式冷凝器 壳管式冷凝器	condenseur à tubes et à calandre	2.5.2-**38**
natural-convection air-cooled condenser	自然对流空冷式冷凝器	condenseur à air statique	2.5.2-**39**
natural-draught cooling tower *atmospheric cooling tower* ○	自然通风冷却塔	refroidisseur d'eau atmosphérique à tirage naturel *tour de refroidissement à tirage naturel*	2.5.2-**40**
open shell-and-tube condenser	立式壳管冷凝器	condenseur à tubes et calandre ouvert	2.5.2-**41**
packing (of a cooling tower) *fill (of a cooling tower)*	冷却塔填料	garnissage (d'un refroidisseur atmosphérique, d'une tour de refroidissement) *remplissage (d'un refroidisseur atmosphérique, d'une tour de refroidissement)* *matelas dispersant*	2.5.2-**42**
plate condenser	板式冷凝器	condenseur à plaques	2.5.2-**43**
plate fill ◐ *plate packing* ◐	片式填料	garnissage à plaques	2.5.2-**44**
plate packing ◐ *plate fill* ◐	片式填料	garnissage à plaques	2.5.2-**45**
random fill ◐ *random packing*	松散填料	garnissage en vrac *remplissage en vrac* ◐	2.5.2-**46**
random packing *random fill* ◐	松散填料	garnissage en vrac *remplissage en vrac* ◐	2.5.2-**47**
Raschig rings	拉西环	anneaux Raschig	2.5.2-**48**
remote mechanical-draft air-cooled refrigerant condenser	外置机械通风式冷凝器	condenseur à air à ventilation forcée autonome	2.5.2-**49**
scale inhibitor	水垢抑制剂	anti-tartre *anti-incrustant* ◐	2.5.2-**50**
shell-and-coil condenser	壳盘管式冷凝器	condenseur à calandre et serpentin *condenseur à virole et serpentin* ◐	2.5.2-**51**
shell-and-tube condenser *(multitubular) shell-and-tube condenser* ◐	壳管式冷凝器 （多管）壳管式冷凝器	condenseur à tubes et à calandre	2.5.2-**52**

	ENGLISH	汉语	FRANÇAIS
2.5.2-**53**	splash packing ◐	飞溅式填料	garnissage dispersant
2.5.2-**54**	split condenser	分体式冷凝器	ensemble multicondenseurs
2.5.2-**55**	spray cooling tower	喷淋式冷却塔	tour de refroidissement à pulvérisation *tour de refroidissement sans garnissage* *refroidisseur d'eau à pulvérisation*
2.5.2-**56**	spray pond	喷水池	refroidisseur d'eau à pulvérisation
2.5.2-**57**	subcooler	过冷器	sous-refroidisseur
2.5.2-**58**	submerged coil condenser ◐ *submerged condenser*	沉浸式冷凝器	condenseur à immersion
2.5.2-**59**	submerged condenser *submerged coil condenser ◐*	沉浸式冷凝器	condenseur à immersion
2.5.2-**60**	tube-in-tube condenser ◐	套管式冷凝器	condenseur à double tube
2.5.2-**61**	warm water *hot water*	热水	eau chaude
2.5.2-**62**	warm water basin *hot water basin*	热水槽	bassin d'accumulation d'eau chaude
2.5.2-**63**	water-cooled condenser	水冷式冷凝器	condenseur à eau
2.5.2-**64**	welded plate condenser ◐	焊接板式冷凝器	condenseur à plaques soudées
2.5.2-**65**	wire-and-tube condenser	绕线式冷凝器	condenseur à tubes et fils

	SECTION 2.5.3 *Evaporators and evaporator accessories*	节 **2.5.3** 蒸发器与蒸发器附件	**SOUS-CHAPITRE 2.5.3** *Evaporateurs et accessoires*
2.5.3-**1**	direct expansion evaporator *dry expansion evaporator*	干式蒸发器	évaporateur à détente sèche
2.5.3-**2**	drier coil	干燥盘管	tube sécheur *échangeur surchauffeur ◐*
2.5.3-**3**	dry expansion evaporator *direct expansion evaporator*	干式蒸发器	évaporateur à détente sèche
2.5.3-**4**	embossed-plate evaporator ○	凹凸板式蒸发器	évaporateur à circuit embouti
2.5.3-**5**	evaporating unit	蒸发机组	groupe évaporateur *groupe d'évaporation ◐*
2.5.3-**6**	evaporator	蒸发器	évaporateur
2.5.3-**7**	flooded evaporator	满液式蒸发器	évaporateur noyé
2.5.3-**8**	forced-air-circulation unit air cooler ◐	强制空气循环式空气冷却器	aérofrigorifère ventilé *aérofrigorifère refroidisseur à convection forcée ◐*
2.5.3-**9**	herringbone-type evaporator ○ *v-coil evaporator*	V型管蒸发器	évaporateur en chevrons *évaporateur en arêtes de poisson ◐*
2.5.3-**10**	ice bank evaporator	蓄冰式蒸发器	évaporateur accumulateur de glace
2.5.3-**11**	integral plate evaporator	整体板式换热器	évaporateur à plaques intégrales
2.5.3-**12**	liquid overfeed evaporator *recirculation-type evaporator*	再循环式蒸发器	évaporateur à recirculation
2.5.3-**13**	liquid suction heat interchanger *superheater ◐*	回热器	échangeur liquide-vapeur
2.5.3-**14**	nominal air flow	名义空气流量	débit d'air nominal

ENGLISH	汉语	FRANÇAIS	
plate coil *tube-on-sheet evaporator* ◐	板管式蒸发器	évaporateur plaque et tubes	2.5.3-**15**
plate evaporator	板式蒸发器	évaporateur platulaire	2.5.3-**16**
pump-fed evaporator	（液体）强制循环式蒸发器	évaporateur noyé alimenté par pompe	2.5.3-**17**
raceway coil ○	平行管蒸发器	évaporateur herse	2.5.3-**18**
recirculation-type evaporator *liquid overfeed evaporator*	再循环式蒸发器	évaporateur à recirculation	2.5.3-**19**
(refrigerant) distributor	（制冷剂）分配器	distributeur *distributeur de frigorigène* ◐	2.5.3-**20**
refrigerant recirculation rate	制冷剂再循环率	taux de circulation	2.5.3-**21**
roll-bond evaporator	压焊式蒸发器	évaporateur "roll bond" *évaporateur soudé par laminage* ◐	2.5.3-**22**
shell-and-coil evaporator	壳盘管式蒸发器	évaporateur à calandre et serpentin	2.5.3-**23**
shell-and-tube evaporator	壳管式蒸发器	évaporateur multitubulaire à calandre *évaporateur à tubes et calandre*	2.5.3-**24**
spray-type evaporator	喷淋式蒸发器	évaporateur à aspersion interne	2.5.3-**25**
superheater ◐ *liquid suction heat interchanger*	回热器	échangeur liquide-vapeur	2.5.3-**26**
tube-on-sheet evaporator ◐ *plate coil*	板管式蒸发器	évaporateur plaque et tubes	2.5.3-**27**
v-coil evaporator *herringbone-type evaporator* ○	V型管蒸发器	évaporateur en chevrons *évaporateur en arêtes de poisson* ◐	2.5.3-**28**
vertical shell-and-tube evaporator	立式壳管式蒸发器	évaporateur multitubulaire vertical	2.5.3-**29**
vertical-type evaporator	立管式蒸发器	évaporateur à tubes verticaux	2.5.3-**30**
wrap-round evaporator	封套式蒸发器	évaporateur à double enveloppe	2.5.3-**31**

SECTION 2.5.4 *Cooling distribution equipment*	节 **2.5.4** 配冷设备	**SOUS-CHAPITRE 2.5.4** *Equipements de distribution de froid*	
ceiling coil *overhead coil* ◐ *ceiling grid* ○	顶盘管	échangeur plafonnier	2.5.4-**1**
ceiling grid ○ *ceiling coil* *overhead coil* ◐	顶盘管	échangeur plafonnier	2.5.4-**2**
coil *grid* ◐	盘管	serpentin *batterie* ◐	2.5.4-**3**
convector (equipment)	对流器（设备）	convecteur	2.5.4-**4**
cooling battery ○	冷却盘管组	refroidisseur d'air *batterie frigorifique* ◐	2.5.4-**5**
cooling coil	冷却盘管	serpentin refroidisseur *élément refroidisseur (tubulaire)* ◐	2.5.4-**6**
eutectic plate	蓄冷板	plaque eutectique	2.5.4-**7**
grid ◐ *coil*	盘管	serpentin *batterie* ◐	2.5.4-**8**
hold-over coil	蓄冷盘管	serpentin accumulateur (de glace)	2.5.4-**9**
hold-over plate	蓄冷板	plaque accumulatrice (de froid)	2.5.4-**10**

	ENGLISH	汉语	FRANÇAIS
2.5.4-**11**	overhead coil ◐ *ceiling coil* *ceiling grid* ○	顶盘管	échangeur plafonnier
2.5.4-**12**	wall coil *wall grid* ◐	墙排管	échangeur mural
2.5.4-**13**	wall grid ◐ *wall coil*	墙排管	échangeur mural

	SECTION 2.6 *Valves and pipes* **SECTION 2.6.1** *Valves: general background*	节 **2.6** 阀与管道 节 **2.6.1** 阀：基础知识	**SOUS-CHAPITRE 2.6** *Robinetterie et tuyauteries* **SOUS-CHAPITRE 2.6.1** *Généralités sur la robinetterie*
2.6.1-**1**	angle pattern body	角阀体	corps de robinet d'équerre
2.6.1-**2**	angle valve	角阀	robinet d'équerre
2.6.1-**3**	ball cock ◐ *ball (float) valve* ◐	浮球阀	robinet à flotteur
2.6.1-**4**	ball (float) valve ◐ *ball cock* ◐	浮球阀	robinet à flotteur
2.6.1-**5**	ball (plug) valve ◐ *plug-and-ball valve*	旋塞球阀 球塞阀	robinet à tournant sphérique *robinet à boule* ◐
2.6.1-**6**	ball valve	弹子阀	clapet à bille
2.6.1-**7**	bellows seal	波纹管密封	soufflet d'étanchéité
2.6.1-**8**	bellows valve	波纹管阀	robinet à soufflet
2.6.1-**9**	body ◐ *valve body*	阀体	corps de robinet *corps* ◐
2.6.1-**10**	body end	阀端	extrémité du corps
2.6.1-**11**	body end port	阀端接口	orifice d'extrémité de corps
2.6.1-**12**	body seat	阀座	siège du corps
2.6.1-**13**	bonnet	阀盖	tête de robinet
2.6.1-**14**	butterfly valve	蝶阀	robinet à papillon
2.6.1-**15**	bypass valve *shunt valve* ◐	旁通阀	robinet de dérivation *robinet de bipasse*
2.6.1-**16**	changeover device	转换装置	dispositif de commutation *inverseur-commutateur*
2.6.1-**17**	charging valve	充注阀	robinet de charge
2.6.1-**18**	check valve *non-return valve* ◐	止回阀	clapet de non-retour *clapet de retenue*
2.6.1-**19**	cock ◐	栓塞阀	robinet (à liquide)
2.6.1-**20**	control valve	调节阀	robinet de régulation *robinet régulateur (piloté)*
2.6.1-**21**	cover	（阀）盖	couvercle
2.6.1-**22**	cover flange	阀盖法兰	bride d'assemblage corps-chapeau ou corps-couvercle
2.6.1-**23**	cylinder valve	气瓶阀	robinet de bouteille
2.6.1-**24**	diaphragm valve *membrane valve*	膜片阀	robinet à membrane

ENGLISH	汉语	FRANÇAIS	
diverting valve	分流阀	robinet diviseur	2.6.1-**25**
double-flanged body	双法兰阀体	corps à brides	2.6.1-**26**
double-seated valve *two-way valve*	双座阀	robinet de service à double siège	2.6.1-**27**
drain hole	放泄孔	purgeur	2.6.1-**28**
drain valve	放泄阀	robinet de soutirage *robinet de vidange*	2.6.1-**29**
effective port area	有效通流面积	section effective de passage	2.6.1-**30**
four-way valve	四通阀	robinet à quatre voies	2.6.1-**31**
gate valve	闸阀	robinet vanne *vanne* ◐ *robinet à vanne* ○	2.6.1-**32**
globe valve	球阀	robinet à soupape *robinet à clapet*	2.6.1-**33**
hand-stop valve	手动截止阀	robinet d'arrêt à main	2.6.1-**34**
initial valve opening	阀的初开度	position minimale d'ouverture	2.6.1-**35**
isolating valve	隔断阀	robinet de sectionnement	2.6.1-**36**
king valve (USA) *master valve*	总（主）阀	robinet général	2.6.1-**37**
lift check valve	升程式止回阀	clapet de retenue à mouvement vertical *clapet de retenue guidé*	2.6.1-**38**
lug-type body	螺栓连接型阀体	corps à oreilles	2.6.1-**39**
mass flow controller	质量流量控制自动阀	clapet d'arrêt	2.6.1-**40**
master valve *king valve (USA)*	总（主）阀	robinet général	2.6.1-**41**
membrane valve *diaphragm valve*	膜片阀	robinet à membrane	2.6.1-**42**
mixing valve	混合阀	robinet mélangeur *robinet mitigeur* ◐	2.6.1-**43**
multi-end body	多接口阀体	corps multivoies	2.6.1-**44**
multiway plug-and-ball valve	多通道球塞阀	robinet à tournant multivoies	2.6.1-**45**
needle valve	针阀	robinet à pointeau *robinet à aiguille*	2.6.1-**46**
non-return valve ◐ *check valve*	止回阀	clapet de non-retour *clapet de retenue*	2.6.1-**47**
oblique pattern body	斜角阀体	corps à tête inclinée	2.6.1-**48**
obturator	闭塞器	obturateur	2.6.1-**49**
oil-charge valve	加油阀	robinet de remplissage d'huile	2.6.1-**50**
oil-drain valve *oil-purge valve* ◐	放油阀	robinet de purge d'huile *robinet d'extraction d'huile*	2.6.1-**51**
oil-purge valve ◐ *oil-drain valve*	放油阀	robinet de purge d'huile *robinet d'extraction d'huile*	2.6.1-**52**
operating mechanism	操作机构	mécanisme de manoeuvre	2.6.1-**53**
packless valve	无填料阀	robinet sans garniture	2.6.1-**54**
permanent bleed-type valve	定量溢流阀	robinet à fuite permanente	2.6.1-**55**

	ENGLISH	汉语	FRANÇAIS
2.6.1-**56**	piston valve *slide valve* ◐	滑阀	robinet à piston
2.6.1-**57**	plug-and-ball valve *ball (plug) valve* ◐	旋塞球阀 球塞阀	robinet à tournant sphérique *robinet à boule* ◐
2.6.1-**58**	plug valve	旋塞阀	robinet à tournant *robinet à boisseau* ○
2.6.1-**59**	port *valve area*	阀断面	section de passage (d'un robinet)
2.6.1-**60**	pump-out valve	抽空阀	robinet de mise sous vide *robinet de mise à vide* ◐
2.6.1-**61**	purge valve	放空阀	robinet de purge *robinet d'extraction* ◐
2.6.1-**62**	refrigerant access valve *Schrader valve*	充注阀	valve de service et de contrôle *valve Schrader*
2.6.1-**63**	refrigerant access valve hose connector	制冷剂充注阀软管接头	flexible de raccordement à la valve Schrader
2.6.1-**64**	refrigerant valve core	制冷剂充注阀芯	noyau d'une valve de service ◐
2.6.1-**65**	regulating valve	调节阀	robinet de réglage *robinet régleur* *régleur*
2.6.1-**66**	reversing valve	换向阀	robinet inverseur
2.6.1-**67**	Schrader valve *refrigerant access valve*	充注阀	valve de service et de contrôle *valve Schrader*
2.6.1-**68**	seating surface	阀座表面	portée d'étanchéité
2.6.1-**69**	service valve	检修阀	robinet de service
2.6.1-**70**	shell	外壳（套）	corps
2.6.1-**71**	shell tapping	外壳（套）壁螺孔	raccordement auxiliaire sur le corps
2.6.1-**72**	shunt valve ◐ *bypass valve*	旁通阀	robinet de dérivation *robinet de bipasse*
2.6.1-**73**	shut-off valve ◐ *stop valve* ◐	截止阀	robinet d'arrêt
2.6.1-**74**	single-flanged body	单法兰阀体	corps monobride
2.6.1-**75**	slide valve ◐ *piston valve*	滑阀	robinet à piston
2.6.1-**76**	stop valve ◐ *shut-off valve* ◐	截止阀	robinet d'arrêt
2.6.1-**77**	straight-pattern body	直通型阀体	corps à tête droite
2.6.1-**78**	swing-check valve	摆式止回阀	clapet de retenue à battant
2.6.1-**79**	tap	丝堵	robinet de puisage
2.6.1-**80**	taper seat	锥形阀座	siège oblique
2.6.1-**81**	three-way valve	三通阀	robinet à trois voies
2.6.1-**82**	throttle valve	节流阀	robinet d'étranglement
2.6.1-**83**	two-way valve *double-seated valve*	双座阀	robinet de service à double siège
2.6.1-**84**	valve area *port*	阀断面	section de passage (d'un robinet)
2.6.1-**85**	valve body *body* ◐	阀体	corps de robinet *corps* ◐
2.6.1-**86**	valve cap	阀盖	chapeau

ENGLISH	汉语	FRANÇAIS	
valve plug	阀体	pointeau	2.6.1-**87**
valve trim	阀内构件	mécanisme du robinet	2.6.1-**88**
valve (in general)	阀	robinet (en général)	2.6.1-**89**
vent hole	泄流孔	évent	2.6.1-**90**

SECTION 2.6.2 *Automatic valves*	节 **2.6.2** 自动阀	**SOUS-CHAPITRE 2.6.2** *Robinets automatiques*	
automatic steam trap	自动疏水器	purgeur automatique de vapeur d'eau	2.6.2-**1**
automatic valve	自动阀	robinet automatique *détendeur automatique* ◐	2.6.2-**2**
back-pressure regulator *evaporator pressure regulator* *back-pressure valve* *constant-pressure valve*	蒸发压力调节阀	vanne de régulation de la pression d'évaporation	2.6.2-**3**
back-pressure valve *evaporator pressure regulator* *back-pressure regulator* *constant-pressure valve*	蒸发压力调节阀	vanne de régulation de la pression d'évaporation	2.6.2-**4**
condensing-pressure valve *high-pressure valve*	冷凝压力阀	vanne de régulation de la pression de condensation	2.6.2-**5**
constant-pressure valve *evaporator pressure regulator* *back-pressure regulator* *back-pressure valve*	蒸发压力调节阀	vanne de régulation de la pression d'évaporation	2.6.2-**6**
evaporator pressure regulator *back-pressure regulator* *back-pressure valve* *constant-pressure valve*	蒸发压力调节阀	vanne de régulation de la pression d'évaporation	2.6.2-**7**
high-pressure valve *condensing-pressure valve*	冷凝压力阀	vanne de régulation de la pression de condensation	2.6.2-**8**
hold-back valve ◐ *suction-pressure regulator*	吸气压力调节阀	vanne de régulation de la pression d'aspiration	2.6.2-**9**
hot gas bypass regulator	热气旁通调节阀	(robinet) régulateur de dérivation des gaz chauds *robinet de bipasse des gaz chauds*	2.6.2-**10**
hydraulically actuated valve	液压阀	robinet à commande hydraulique	2.6.2-**11**
liquid-injection valve	喷液阀	robinet d'injection de liquide *détendeur d'injection*	2.6.2-**12**
magnetic valve *solenoid valve*	电磁阀	robinet électromagnétique *robinet solénoïde* *électrovanne* ◐ *robinet d'arrêt électromagnétique* ◐	2.6.2-**13**
modulating valve	调节阀	robinet modulant	2.6.2-**14**
motor-operated valve *motorized valve* ◐	电动阀	robinet motorisé	2.6.2-**15**

	ENGLISH	汉语	FRANÇAIS
2.6.2-**16**	motorized valve ◐ *motor-operated valve*	电动阀	robinet motorisé
2.6.2-**17**	pilot-controlled valve *servo-operated valve ◐*	伺服阀	robinet à servocommande
2.6.2-**18**	pilot valve	导阀	robinet pilote
2.6.2-**19**	pneumatic-operated valve	气动阀	robinet pneumatique
2.6.2-**20**	pressure control valve	压力控制阀	vanne de régulation de pression
2.6.2-**21**	pressure-controlled valve	压力调节阀	robinet pressostatique
2.6.2-**22**	(pressure) reducing valve	减压阀	robinet réducteur de pression
2.6.2-**23**	quick-closing valve	快速关闭阀 快关阀	robinet à fermeture rapide *soupape à fermeture rapide*
2.6.2-**24**	servo-operated valve ◐ *pilot-controlled valve*	伺服阀	robinet à servocommande
2.6.2-**25**	snap-action valve	速动阀	régulateur de pression à déclic
2.6.2-**26**	solenoid valve *magnetic valve*	电磁阀	robinet électromagnétique *robinet solénoïde* *électrovanne ◐* *robinet d'arrêt électromagnétique ◐*
2.6.2-**27**	suction-pressure regulator *hold-back valve ◐*	吸气压力调节阀	vanne de régulation de la pression d'aspiration
2.6.2-**28**	thermostatically controlled valve	热力调节阀	robinet thermostatique
2.6.2-**29**	water-regulating valve	水调节阀	(robinet) régulateur de débit d'eau

	SECTION 2.6.3 *Expansion devices*	节 **2.6.3** 膨胀装置	**SOUS-CHAPITRE 2.6.3** *Détendeurs*
2.6.3-**1**	adjustable capillary valve ○	可调毛细通道阀	robinet capillaire réglable
2.6.3-**2**	capacity of an expansion valve	膨胀阀容量	capacité frigorifique d'un détendeur *puissance frigorifique d'un détendeur ◐*
2.6.3-**3**	capillary tube	毛细管	capillaire
2.6.3-**4**	compensator	补偿器	dispositif d'affranchissement de l'ambiance
2.6.3-**5**	direct-acting valve	直接作用阀	détendeur thermostatique à action directe
2.6.3-**6**	expansion device	膨胀机构	dispositif de détente
2.6.3-**7**	expansion valve	膨胀阀	détendeur *robinet détendeur ◐*
2.6.3-**8**	external equalizer	外平衡管	égalisateur externe
2.6.3-**9**	float-type expansion valve ○ *float valve ◐*	浮球阀	régleur à flotteur ◐ *détendeur à flotteur ○*
2.6.3-**10**	float valve ◐ *float-type expansion valve ○*	浮球阀	régleur à flotteur ◐ *détendeur à flotteur ○*
2.6.3-**11**	gas-charged thermostat *limited liquid charged thermostat*	充气式恒温器 有限充液恒温器	thermostat à charge (liquide) limitée *thermostat chargé en vapeur*
2.6.3-**12**	hand expansion valve	手动膨胀阀	détendeur à main
2.6.3-**13**	high-pressure float valve *high-side float valve (USA) ◐*	高压浮球阀	détendeur à flotteur haute pression *régleur à flotteur haute pression ◐*

ENGLISH	汉语	FRANÇAIS	
high-side float valve (USA) ◐ *high-pressure float valve*	高压浮球阀	détendeur à flotteur haute pression *régleur à flotteur haute pression* ◐	2.6.3-**14**
internal equalizer	内平衡管	égalisateur interne	2.6.3-**15**
limited liquid charged thermostat *gas-charged thermostat*	充气式恒温器 有限充液恒温器	thermostat à charge (liquide) limitée *thermostat chargé en vapeur*	2.6.3-**16**
liquid-charged thermostat	充液恒温器	thermostat chargé en liquide *thermostat chargé en vapeur saturée-liquide* *thermostat à charge liquide*	2.6.3-**17**
low-pressure float valve *low-side float valve (USA)* ◐	低压浮球阀	régleur à flotteur basse pression	2.6.3-**18**
low-side float valve (USA) ◐ *low-pressure float valve*	低压浮球阀	régleur à flotteur basse pression	2.6.3-**19**
maximum operating pressure (MOP)	最大工作压力	pression maximale de fonctionnement	2.6.3-**20**
multiport expansion valve	多接口膨胀阀	détendeur à orifices multiples *détendeur multi-orifices*	2.6.3-**21**
operating superheat	运行过热	surchauffe de fonctionnement *surchauffe opérationnelle*	2.6.3-**22**
pilot-operated expansion valve	组合阀	(robinet) détendeur piloté	2.6.3-**23**
power element (of a thermostat) *power system (of a thermostat)* ◐	（恒温器的）驱动部件	train thermostatique *actionneur*	2.6.3-**24**
power system (of a thermostat) ◐ *power element (of a thermostat)*	（恒温器的）驱动部件	train thermostatique *actionneur*	2.6.3-**25**
refrigerant metering device	制冷剂流量调节机构	appareil de réglage de débit de fluide *régleur*	2.6.3-**26**
restrictor	（压差）限制器	restricteur	2.6.3-**27**
restrictor valve	限制器阀	dispositif restricto-changeur	2.6.3-**28**
static superheat	静态过热	seuil de surchauffe	2.6.3-**29**
superheat change	过热变化	surchauffe effective	2.6.3-**30**
temperature-sensing element	感温元件	bulbe du train thermostatique	2.6.3-**31**
(thermostatic) adsorber charge	吸附充注	charge adsorbante (thermostatique)	2.6.3-**32**
(thermostatic) cross charge	异工质充注	hétérocharge (thermostatique)	2.6.3-**33**
(thermostatic) dry charge ◐ *(thermostatic) gas charge* *(thermostatic) limited liquid charge* ◐	气体充注	charge limitée (thermostatique)	2.6.3-**34**
thermostatic expansion valve	热力膨胀阀	détendeur thermostatique	2.6.3-**35**
thermostatic expansion valve charge	热力膨胀阀充注	charge de détendeur thermostatique	2.6.3-**36**
thermostatic expansion valve pressure drop	热力膨胀阀压力降	chute de pression à travers le détendeur	2.6.3-**37**
thermostatic expansion valve superheat	热力膨胀阀过热度	surchauffe d'un détendeur thermo-statique	2.6.3-**38**
(thermostatic) gas charge *(thermostatic) dry charge* ◐ *(thermostatic) limited liquid charge* ◐	气体充注	charge limitée (thermostatique)	2.6.3-**39**
(thermostatic) limited liquid charge ◐ *(thermostatic) gas charge* *(thermostatic) dry charge* ◐	气体充注	charge limitée (thermostatique)	2.6.3-**40**
(thermostatic) liquid charge *(thermostatic) wet charge* ◐	液体充注	charge liquide (thermostatique)	2.6.3-**41**

	ENGLISH	汉语	FRANÇAIS
2.6.3-**42**	(thermostatic) straight charge ◐	同工质充注	homocharge (thermostatique)
2.6.3-**43**	(thermostatic) wet charge ◐ *(thermostatic) liquid charge*	液体充注	charge liquide (thermostatique)

	SECTION 2.6.4 *Piping and fittings*	节 **2.6.4** 管道与附件	**SOUS-CHAPITRE 2.6.4** *Tuyauteries et raccords*
2.6.4-**1**	braze-welding	铜焊	soudobrasage
2.6.4-**2**	brazing	铜焊	brasage
2.6.4-**3**	brine line	盐水管	conduite de saumure
2.6.4-**4**	butt welding end	对焊端	extrémité à souder en bout
2.6.4-**5**	bypass	旁通管	bipasse *dérivation* ◐
2.6.4-**6**	cap	管帽	bouchon femelle *capuchon*
2.6.4-**7**	capillary end	钎焊或锡焊端	extrémité à braser par capillarité
2.6.4-**8**	capillary fitting *sweat joint* ◐	钎焊接头	joint brasé par capillarité
2.6.4-**9**	charging connection	充注接头	raccord de charge
2.6.4-**10**	clamp ring	紧固圈	collier de serrage
2.6.4-**11**	coil of tubing	铜盘管	couronne de tube
2.6.4-**12**	condensate line	冷凝液管	conduite de condensat
2.6.4-**13**	connecting hose *hose assembly* ◐	软接头	flexible (de raccordement)
2.6.4-**14**	coupling	接管	manchon (de raccordement)
2.6.4-**15**	discharge line *hot gas line* ◐	排气管	conduite de refoulement *tuyauterie de refoulement*
2.6.4-**16**	double (suction) riser	双升管	conduite ascendante double
2.6.4-**17**	double male reduction	异径外承插管	réduction double mâle
2.6.4-**18**	expansion bend	伸缩弯	lyre de dilatation
2.6.4-**19**	expansion joint	伸缩节	joint de dilatation
2.6.4-**20**	expansion loop ◐	伸缩管	boucle de dilatation *cor de chasse de dilatation*
2.6.4-**21**	fittings	管件	raccords
2.6.4-**22**	flange	法兰	bride
2.6.4-**23**	flanged end	法兰端面	extrémité à bride
2.6.4-**24**	flanged joint	法兰接头	joint à bride
2.6.4-**25**	flare fitting *flared joint*	喇叭口接头	collet *joint conique* *joint évasé* *"dudgeon"* ◐
2.6.4-**26**	flared joint *flare fitting*	喇叭口接头	collet *joint conique* *joint évasé* *"dudgeon"* ◐

ENGLISH	汉语	FRANÇAIS	
flaring block ◐ *flaring tool*	扩口工具	dudgeonnière *outil à façonner les collets*	2.6.4-**27**
flaring tool *flaring block* ◐	扩口工具	dudgeonnière *outil à façonner les collets*	2.6.4-**28**
flexible pipe element	软管构件	élément flexible de tuyauterie	2.6.4-**29**
gasket	垫	joint (matériau) *garniture* ◐	2.6.4-**30**
hard soldering	硬焊	brasage fort	2.6.4-**31**
header *manifold*	集管 歧管	collecteur (à raccordements multiples) *manifold* ◐	2.6.4-**32**
heavy gauge	厚规格	de forte épaisseur	2.6.4-**33**
hose assembly ◐ *connecting hose*	软接头	flexible (de raccordement)	2.6.4-**34**
hot gas defrost line	除霜热气管	conduite de gaz chaud pour dégivrage	2.6.4-**35**
hot gas line ◐ *discharge line*	排气管	conduite de refoulement *tuyauterie de refoulement*	2.6.4-**36**
joint (in general)	连接（总称）	joint (en général)	2.6.4-**37**
joint ring ○	密封垫圈	joint torique *joint annulaire* ◐	2.6.4-**38**
line *main* ◐	管线	canalisation *conduite* *tuyauterie* ◐	2.6.4-**39**
liquid line	液体管	tuyauterie de liquide *conduite de liquide* ◐	2.6.4-**40**
main ◐ *line*	管线	canalisation *conduite* *tuyauterie* ◐	2.6.4-**41**
main pipe	主管	canalisation principale *conduite principale*	2.6.4-**42**
male-female facing flange *raised-face flange* ◐ *R-F flange* ◐	凹凸面法兰	bride à emboîtement simple *bride à face surélevée*	2.6.4-**43**
manifold *header*	集管 歧管	collecteur (à raccordements multiples) *manifold* ◐	2.6.4-**44**
mechanical joint	机械连接接头	joint mécanique	2.6.4-**45**
metallic flexible pipe	金属软管	tuyauterie flexible métallique	2.6.4-**46**
nipple	螺纹接管	mamelon	2.6.4-**47**
O-ring joint	0型垫圈	joint torique (à section circulaire)	2.6.4-**48**
pipe schedule	管道明细表	spécification des tubes	2.6.4-**49**
piping	配管	tuyauterie	2.6.4-**50**
plug	塞子	bouchon mâle	2.6.4-**51**
quick-coupling *quick-release coupling* ◐	快装接头	raccord instantané *raccord rapide*	2.6.4-**52**
quick-release coupling ◐ *quick-coupling*	快装接头	raccord instantané *raccord rapide*	2.6.4-**53**
R-F flange ◐ *male-female facing flange* *raised-face flange* ◐	凹凸面法兰	bride à emboîtement simple *bride à face surélevée*	2.6.4-**54**

	ENGLISH	汉语	FRANÇAIS
2.6.4-**55**	raised-face flange ◐ *male-female facing flange* *R-F flange* ◐	凹凸面法兰	bride à emboîtement simple *bride à face surélevée*
2.6.4-**56**	reducing bushing	缩径套管	réduction mâle-femelle
2.6.4-**57**	reducing coupling	异径内承插管	réduction femelle-femelle *réduction double femelle*
2.6.4-**58**	return line ◐ *suction line*	吸气管	tuyauterie d'aspiration *canalisation d'aspiration* ◐ *conduite d'aspiration* ◐
2.6.4-**59**	screwed joint	螺纹接头	joint vissé
2.6.4-**60**	sealant	密封胶	produits d'étanchéité
2.6.4-**61**	seam	接缝	couture
2.6.4-**62**	seamless pipe	无缝管	tube (étiré) sans soudure
2.6.4-**63**	socket female end	承插口	extrémité à emboîter femelle *embout femelle*
2.6.4-**64**	socket welding end	承插焊口	extrémité à emboîter et à souder *embout à souder*
2.6.4-**65**	(soft) soldering	软焊	brasage tendre
2.6.4-**66**	soldered joint	软焊接头	joint brasé tendre
2.6.4-**67**	spool piece	双端凸缘管	entretoise à brides *raccord à brides* ◐ *manchette* ◐
2.6.4-**68**	suction line *return line* ◐	吸气管	tuyauterie d'aspiration *canalisation d'aspiration* ◐ *conduite d'aspiration* ◐
2.6.4-**69**	swaging tool	胀口工具	appareil à façonner les emboîtures *appareil à mandriner* ◐
2.6.4-**70**	sweat joint ◐ *capillary fitting*	钎焊接头	joint brasé par capillarité
2.6.4-**71**	taper pipe thread end	锥形管螺纹端	joint fileté conique
2.6.4-**72**	threaded end	螺纹端	extrémité filetée
2.6.4-**73**	tongue-and-groove facing flange ◐	榫槽面法兰	bride à emboîtement double
2.6.4-**74**	vibration isolator	隔振器	isolateur de vibration
2.6.4-**75**	weld neck flange	高颈法兰	bride à collerette (à souder en bout)
2.6.4-**76**	welded joint	焊接接头	joint soudé
2.6.4-**77**	welded tube	焊接管	tube soudé
2.6.4-**78**	welding	焊接	soudage autogène
2.6.4-**79**	welding end	焊接端	embout à souder

	SECTION 2.7 *Other equipment used for refrigeration production*	节 **2.7** 用于制冷的其它设备	**SOUS-CHAPITRE 2.7** *Autres équipements utilisés pour la production de froid*
2.7-**1**	accumulator	低压(循环)桶	réservoir tampon *bouteille tampon* ◐
2.7-**2**	back-pressure gauge ○ *low-pressure gauge* *suction gauge*	低压表 吸气压力表	manomètre basse pression (B.P.) *manomètre d'aspiration*

ENGLISH	汉语	FRANÇAIS	
balance tank	平衡罐	réservoir d'équilibre	2.7-**3**
buffer tank	缓冲容器	réservoir tampon	2.7-**4**
centrifugal pump	离心泵	pompe centrifuge	2.7-**5**
chilled water jacketed tank	冷水夹套罐	bac à enveloppe d'eau glacée	2.7-**6**
circulation pump	循环泵	pompe de circulation	2.7-**7**
compartment	隔舱	compartiment	2.7-**8**
component	组件	composant	2.7-**9**
dead end trap	终端集液器	piège à l'aspiration	2.7-**10**
dehydrator *drier*	干燥器	déshydrateur *dessiccateur* ◐ *sécheur* ◐	2.7-**11**
desiccant	干燥剂	déshydratant	2.7-**12**
discharge gauge (UK) discharge gage (USA) *high-pressure gauge* *head-pressure gauge* ◐	排气压力表 高压表	manomètre de refoulement *manomètre haute pression*	2.7-**13**
drier *dehydrator*	干燥器	déshydrateur *dessiccateur* ◐ *sécheur* ◐	2.7-**14**
dump trap liquid return	液体回收	retour de liquide à la haute pression	2.7-**15**
equalizer	均压管	égalisateur (de pression)	2.7-**16**
equalizer tank	均压罐	réservoir égalisateur	2.7-**17**
filter *strainer* ◐ *screen* ○	过滤器	filtre	2.7-**18**
filter-dehydrator ◐ *filter-drier*	干燥-过滤器	déshydrateur-filtre	2.7-**19**
filter-drier *filter-dehydrator* ◐	干燥-过滤器	déshydrateur-filtre	2.7-**20**
flash chamber	闪发气体分离器	chambre de séparation (après détendeur)	2.7-**21**
flash intercooler	闪发式中间冷却器	refroidisseur intermédiaire à détente	2.7-**22**
frost level indicator	结霜液位指示器	niveau à givrage	2.7-**23**
gas purger ◐ *non-condensable gas purger*	不凝性气体分离器 放空气器	désaérateur *purgeur de gaz non condensables* *dégazeur* ◐	2.7-**24**
gear pump	齿轮泵	pompe à engrenage	2.7-**25**
head-pressure gauge ◐ *discharge gauge (UK) discharge gage (USA)* *high-pressure gauge*	排气压力表 高压表	manomètre de refoulement *manomètre haute pression*	2.7-**26**
high-pressure gauge *discharge gauge (UK) discharge gage (USA)* *head-pressure gauge* ◐	排气压力表 高压表	manomètre de refoulement *manomètre haute pression*	2.7-**27**
high-side receiver *receiver* *liquid receiver*	贮液器	réservoir de liquide haute pression *bouteille accumulatrice de liquide haute pression*	2.7-**28**
intercooler *interstage cooler* ◐	中间冷却器	refroidisseur intermédiaire	2.7-**29**
interstage cooler ◐ *intercooler*	中间冷却器	refroidisseur intermédiaire	2.7-**30**

	ENGLISH	汉语	FRANÇAIS
2.7-**31**	liquid flow indicator	液流指示器	voyant (de) liquide *contrôleur de circulation* *indicateur de passage de liquide*
2.7-**32**	liquid level indicator	液位指示器	indicateur de niveau (de liquide)
2.7-**33**	liquid pocket	液囊	poche de liquide
2.7-**34**	liquid receiver *receiver* *high-side receiver*	贮液器	réservoir de liquide haute pression *bouteille accumulatrice de liquide haute pression*
2.7-**35**	liquid separator ◐ *suction accumulator* ◐ *suction trap* ◐	液体分离器 气液分离器	séparateur de liquide *bouteille anti-coup de liquide*
2.7-**36**	liquid trap	集液器	piège à liquide
2.7-**37**	low-pressure gauge *suction gauge* *back-pressure gauge* ○	低压表 吸气压力表	manomètre basse pression (B.P.) *manomètre d'aspiration*
2.7-**38**	low-pressure receiver *surge drum* ◐ *surge tank* ◐	低压贮液桶	réservoir de liquide basse pression *ballon basse pression*
2.7-**39**	non-condensable gas purger *gas purger* ◐	不凝性气体分离器 放空气器	désaérateur *purgeur de gaz non condensables* *dégazeur* ◐
2.7-**40**	oil drain	放油管	purgeur d'huile
2.7-**41**	oil pocket *oil slug*	油囊	poche d'huile
2.7-**42**	oil receiver	集油器	récepteur d'huile
2.7-**43**	oil rectifier *oil still* ◐	油处理器	rectificateur d'huile
2.7-**44**	oil separator	油分离器	séparateur d'huile *déshuileur*
2.7-**45**	oil slug *oil pocket*	油囊	poche d'huile
2.7-**46**	oil still ◐ *oil rectifier*	油处理器	rectificateur d'huile
2.7-**47**	oil trap	集油器	piège à huile
2.7-**48**	pressure equipment	耐压设备	équipement sous pression
2.7-**49**	pressure vessel	压力容器	récipient sous pression
2.7-**50**	purge recovery system *purge unit* *purging device*	净化系统 放气装置	groupe de purge *florentin* ◐
2.7-**51**	purge unit *purge recovery system* *purging device*	净化系统 放气装置	groupe de purge *florentin* ◐
2.7-**52**	purging device *purge recovery system* *purge unit*	净化系统 放气装置	groupe de purge *florentin* ◐
2.7-**53**	receiver *high-side receiver* *liquid receiver*	贮液器	réservoir de liquide haute pression *bouteille accumulatrice de liquide haute pression*
2.7-**54**	scale trap	杂质沉淀器	séparateur d'impuretés

ENGLISH	汉语	FRANÇAIS	
screen ○ *filter* *strainer* ◐	过滤器	filtre	2.7-**55**
sight glass	视镜	voyant *regard*	2.7-**56**
strainer ◐ *filter* *screen* ○	过滤器	filtre	2.7-**57**
suction accumulator ◐ *liquid separator* ◐ *suction trap* ◐	液体分离器 气液分离器	séparateur de liquide *bouteille anti-coup de liquide*	2.7-**58**
suction gauge *low-pressure gauge* *back-pressure gauge* ○	低压表 吸气压力表	manomètre basse pression (B.P.) *manomètre d'aspiration*	2.7-**59**
suction trap ◐ *liquid separator* ◐ *suction accumulator* ◐	液体分离器 气液分离器	séparateur de liquide *bouteille anti-coup de liquide*	2.7-**60**
surge drum ◐ *low-pressure receiver* *surge tank* ◐	低压贮液桶	réservoir de liquide basse pression *ballon basse pression*	2.7-**61**
surge tank ◐ *low-pressure receiver* *surge drum* ◐	低压贮液桶	réservoir de liquide basse pression *ballon basse pression*	2.7-**62**

SECTION 2.8 *Control – Safety devices*	节 **2.8** 控制——安全装置	**SOUS-CHAPITRE 2.8** *Régulation – Contrôle – Sécurité*	
actuator	执行机构	actionneur	2.8-**1**
adjustability	调整范围	plage de réglage *possibilité de réglage* ◐	2.8-**2**
adjustment	调整	réglage *mise au point* ◐	2.8-**3**
automatic control	自动控制	commande automatique	2.8-**4**
automatic control device	自动控制装置	appareil de régulation automatique	2.8-**5**
automatic control engineering	自控工程	automatique (subst.)	2.8-**6**
automatic operation	自动操作	automatisme	2.8-**7**
automatic sequence control ◐	自动程序控制	automatisme séquentiel	2.8-**8**
automation	自动化	automatisation	2.8-**9**
bursting disc *rupture disc* *safety disc* ◐ *frangible disc* ○	安全片	disque de rupture *disque de sûreté* *membrane d'éclatement*	2.8-**10**
cascade control system	串级控制系统	régulation en cascade	2.8-**11**
change-over switch *selector switch* ◐	转换开关	commutateur	2.8-**12**
closed loop *feedback loop*	闭环	boucle de régulation	2.8-**13**
control	控制	commande *régulation (en général)*	2.8-**14**

	ENGLISH	汉语	FRANÇAIS
2.8-**15**	control console	控制台	pupitre de commande *console de commande* ◐
2.8-**16**	control panel	控制屏	armoire de commande *coffret de commande* ◐ *tableau de commande* ◐
2.8-**17**	control rate	调节速率	vitesse de régulation *délai de réponse* *temps de réponse*
2.8-**18**	control thermostat	控制恒温器	thermostat de régulation
2.8-**19**	controller	控制器	régulateur *dispositif de commande* ◐ *dispositif de régulation* ◐
2.8-**20**	counterweight safety valve	配重安全阀	soupape de sûreté à contrepoids
2.8-**21**	cut-in-point	接通点	point d'enclenchement
2.8-**22**	cut-out point	断开点	point de coupure *point de déclenchement*
2.8-**23**	damping	阻尼	amortissement
2.8-**24**	dead band *dead zone* ◐	死区	zone d'insensibilité *plage d'insensibilité* *zone morte*
2.8-**25**	dead time *lag time*	滞后时间	temps mort
2.8-**26**	dead zone ◐ *dead band*	死区	zone d'insensibilité *plage d'insensibilité* *zone morte*
2.8-**27**	detecting element *sensor* *sensing element* ◐	传感器 检测元件 敏感元件	capteur *élément sensible*
2.8-**28**	differential controller	差动控制器	régulateur différentiel *appareil de régulation à différentiel* ◐
2.8-**29**	differential of a controller	控制器的差值	différentiel d'un régulateur *fourchette d'un régulateur*
2.8-**30**	dual pressure controller	高低压控制器	pressostat combiné haute pression-basse pression
2.8-**31**	feedback control	反馈控制	commande par rétroaction
2.8-**32**	feedback loop *closed loop*	闭环	boucle de régulation
2.8-**33**	feeler (bulb) ○ *thermostat bulb* *sensing bulb* ◐ *thermostat vial* ○	恒温器温包	bulbe (sensible) *bulbe (thermostatique)*
2.8-**34**	float switch	浮子开关	interrupteur à flotteur
2.8-**35**	frangible disc ○ *bursting disc* *rupture disc* *safety disc* ◐	安全片	disque de rupture *disque de sûreté* *membrane d'éclatement*
2.8-**36**	fusible plug	易熔塞	fusible *bouchon fusible*
2.8-**37**	graduated acting	分级动作	modulant *à action progressive*
2.8-**38**	halide torch	卤素灯	lampe haloïde

ENGLISH	汉语	FRANÇAIS	
high discharge temperature cut-out	排气温度过高切断器	thermostat de sécurité de refoulement	2.8-**39**
high-low action ◐ *high-low control*	高-低控制	commande par tout ou peu	2.8-**40**
high-low control *high-low action* ◐	高-低控制	commande par tout ou peu	2.8-**41**
high-pressure controller	高压控制器	pressostat haute pression	2.8-**42**
high-pressure safety cut-out	高压安全切断器	pressostat de sécurité haute pression	2.8-**43**
humidistat *hygrostat* ◐	恒湿器	hygrostat *humidostat* ○	2.8-**44**
humidity controller	湿度控制器	régulateur d'humidité *dispositif de réglage d'humidité* ◐	2.8-**45**
hunting	摆动	pompage	2.8-**46**
hygrostat ◐ *humidistat*	恒湿器	hygrostat *humidostat* ○	2.8-**47**
inherent regulation	自平衡调节	autorégulation	2.8-**48**
integral action controller	整体动作控制器	régulateur intégral	2.8-**49**
intrinsic pressure safety	内在压力安全性	sécurité intrinsèque pour la pression	2.8-**50**
lag time *dead time*	滞后时间	temps mort	2.8-**51**
leak detection	检漏	détection des fuites	2.8-**52**
leak detector	检漏器	détecteur de fuites	2.8-**53**
leak rate	泄漏率	débit de fuite	2.8-**54**
leakage test pressure	泄漏测试压力	pression d'essai de fuite *pression de l'essai de fuite*	2.8-**55**
low-pressure controller	低压控制器	pressostat basse pression	2.8-**56**
low-pressure safety cut-out *suction pressure safety cut-out*	低压安全切断器 吸气压力安全切断	pressostat de sécurité basse pression	2.8-**57**
low-suction temperature cut-out	吸气温度切断器	thermostat de sécurité d'aspiration	2.8-**58**
master controller	主令控制器	régulateur principal	2.8-**59**
measuring unit	测量装置	transmetteur (de mesure)	2.8-**60**
mercury switch	水银开关	basculeur à mercure *interrupteur à mercure*	2.8-**61**
modulating control	幅度控制	régulation modulante	2.8-**62**
monitoring	监控	contrôle automatique *monitorage* ◐	2.8-**63**
oil-charging pump	注油泵	pompe de chargement d'huile	2.8-**64**
oil failure switch ○ *oil pressure cut-out* *oil pressure switch* ◐	油压切断器	pressostat de sécurité d'huile	2.8-**65**
oil pressure cut-out *oil pressure switch* ◐ *oil failure switch* ○	油压切断器	pressostat de sécurité d'huile	2.8-**66**
oil pressure switch ◐ *oil pressure cut-out* *oil failure switch* ○	油压切断器	pressostat de sécurité d'huile	2.8-**67**
oil temperature cut-out	油温切断器	thermostat de sécurité d'huile	2.8-**68**

	ENGLISH	汉语	FRANÇAIS
2.8-**69**	on-off action ◐ *on-off control* *two-step control*	开-关控制	commande par tout ou rien *action par tout ou rien* ◐
2.8-**70**	on-off control *two-step control* *on-off action* ◐	开-关控制	commande par tout ou rien *action par tout ou rien* ◐
2.8-**71**	positioner	定位器	positionneur
2.8-**72**	power-assisted control	外力控制	régulation indirecte
2.8-**73**	pressostat ◐ *pressure switch* ◐	压力继电器 压力开关	pressostat
2.8-**74**	pressure controller	压力控制器	régulateur de pression *dispositif de réglage de pression* ◐
2.8-**75**	pressure cut-out *pressure-limiting device with manual reset* *pressure-limiting device with safety manual reset*	压力切断器 手动复位压力限制装置 安全手动复位的压力限制装置	dispositif de limitation de pression avec réarmement manuel *dispositif de limitation de pression avec réarmement manuel de sécurité*
2.8-**76**	pressure differential cut-out	压差控制器	pressostat différentiel
2.8-**77**	pressure limiter *pressure-limiting device with automatic reset*	压力限制器 自动复位压力限制装置	limiteur de pression automatique *dispositif de limitation de pression avec réarmement automatique* *pressostat haute ou basse pression à réenclenchement automatique*
2.8-**78**	pressure-limiting device ◐ pressure-relief device ◐	压力限制装置 泄压装置	limiteur de pression
2.8-**79**	pressure-limiting device with automatic reset *pressure limiter*	压力限制器 自动复位压力限制装置	limiteur de pression automatique *dispositif de limitation de pression avec réarmement automatique* *pressostat haute ou basse pression à réenclenchement automatique*
2.8-**80**	pressure-limiting device with manual reset *pressure cut-out* *pressure-limiting device with safety manual reset*	压力切断器 手动复位压力限制装置 安全手动复位的压力限制装置	dispositif de limitation de pression avec réarmement manuel *dispositif de limitation de pression avec réarmement manuel de sécurité*
2.8-**81**	pressure-limiting device with safety manual reset *pressure cut-out* *pressure-limiting device with manual reset*	压力切断器 手动复位压力限制装置 安全手动复位的压力限制装置	dispositif de limitation de pression avec réarmement manuel *dispositif de limitation de pression avec réarmement manuel de sécurité*
2.8-**82**	pressure-limiting type valve	压力限制型阀门	détendeur à pression maximale d'ouverture (MOP) *détendeur à charge limitée*
2.8-**83**	pressure-relief device ◐ pressure-limiting device ◐	压力限制装置 泄压装置	limiteur de pression
2.8-**84**	pressure-relief valve *safety valve* ◐	安全阀	soupape de sûreté *soupape limiteur de pression* ◐
2.8-**85**	pressure switch ◐ *pressostat* ◐	压力继电器 压力开关	pressostat
2.8-**86**	proportional-action controller	比例作用控制器	régulateur proportionnel *régulateur à simple action*
2.8-**87**	Proportional Integral Derivative (PID) controller	比例积分微分（PID）控制器	régulateur PID

ENGLISH	汉语	FRANÇAIS	
quick-closing valve	快速关闭阀 快关阀	robinet à fermeture rapide *soupape à fermeture rapide*	2.8-**88**
refrigerant detector	制冷剂检测器	détecteur de fluide frigorigène	2.8-**89**
relay	继电器	relais	2.8-**90**
remote bulb thermostat	遥感恒温器	thermostat à bulbe (et capillaire)	2.8-**91**
remote control	遥控	télécommande *commande à distance* ◐	2.8-**92**
response time	反应时间	temps de réponse	2.8-**93**
room thermostat	室用恒温器	thermostat d'ambiance	2.8-**94**
rupture disc *bursting disc* *safety disc* ◐ *frangible disc* ○	安全片	disque de rupture *disque de sûreté* *membrane d'éclatement*	2.8-**95**
rupture member	破裂构件	dispositif de rupture *élément de rupture*	2.8-**96**
safety cut-out	安全切断器	dispositif de sécurité par coupure	2.8-**97**
safety device	安全装置	dispositif de sécurité *organe de sécurité*	2.8-**98**
safety disc ◐ *bursting disc* *rupture disc* *frangible disc* ○	安全片	disque de rupture *disque de sûreté* *membrane d'éclatement*	2.8-**99**
safety pressure cut-out	安全压力切断器	pressostat de sécurité	2.8-**100**
safety valve ◐ *pressure-relief valve*	安全阀	soupape de sûreté *soupape limiteur de pression* ◐	2.8-**101**
selector switch ◐ *change-over switch*	转换开关	commutateur	2.8-**102**
self-contained breathing apparatus	自给式呼吸装置	appareil de protection respiratoire autonome	2.8-**103**
self-operated control ◐	自力控制	régulation directe	2.8-**104**
self-operated measuring unit ◐	直接测量装置	transmetteur direct	2.8-**105**
sensing bulb ◐ *thermostat bulb* *feeler (bulb)* ○ *thermostat vial* ○	恒温器温包	bulbe (sensible) *bulbe (thermostatique)*	2.8-**106**
sensing element ◐ *sensor* *detecting element*	传感器 检测元件 敏感元件	capteur *élément sensible*	2.8-**107**
sensor *detecting element* *sensing element* ◐	传感器 检测元件 敏感元件	capteur *élément sensible*	2.8-**108**
servocontrol	伺服控制	servocommande	2.8-**109**
set point (of a controller) *set value (of a controller)*	（控制器的）设定值	point de consigne *point de réglage* *valeur de consigne* ○ *valeur de réglage* ○	2.8-**110**
set value (of a controller) *set point (of a controller)*	（控制器的）设定值	point de consigne *point de réglage* *valeur de consigne* ○ *valeur de réglage* ○	2.8-**111**
shut-off device	关闭装置	dispositif d'arrêt	2.8-**112**

	ENGLISH	汉语	FRANÇAIS
2.8-**113**	solid-state device	固态器件	appareil transistorisé
2.8-**114**	spring-loaded pressure-relief valve	弹簧式安全阀	soupape de sûreté à ressort
2.8-**115**	step control *step-by-step control* ○	步进控制	commande pas-à-pas
2.8-**116**	step-by-step control ○ *step control*	步进控制	commande pas-à-pas
2.8-**117**	step controller	步进控制器	"step controller" *dispositif de commande pas-à-pas* ◐ *dispositif de réglage pas-à-pas* ◐
2.8-**118**	strength-test pressure	强度测试压力 试验压力	pression de l'essai de résistance *pression d'épreuve* ◐
2.8-**119**	submaster	辅助控制器	régulateur secondaire
2.8-**120**	suction pressure safety cut-out *low-pressure safety cut-out*	低压安全切断器 吸气压力安全切断	pressostat de sécurité basse pression
2.8-**121**	temperature controller	温度控制器	régulateur de température *dispositif de réglage de température* ◐
2.8-**122**	temperature-limiting device	温度限制装置	dispositif de limitation de la température
2.8-**123**	thermostat	恒温器	thermostat
2.8-**124**	thermostat bulb *sensing bulb* ◐ *feeler (bulb)* ○ *thermostat vial* ○	恒温器温包	bulbe (sensible) *bulbe (thermostatique)*
2.8-**125**	thermostat vial ○ *thermostat bulb* *sensing bulb* ◐ *feeler (bulb)* ○	恒温器温包	bulbe (sensible) *bulbe (thermostatique)*
2.8-**126**	timer	定时器	chronorelais *minuteur* *relais chronométrique*
2.8-**127**	transmitter	传感器	transmetteur (indirect)
2.8-**128**	two-step control *on-off control* *on-off action* ◐	开-关控制	commande par tout ou rien *action par tout ou rien* ◐

	SECTION 2.9 *Working fluids* **SECTION 2.9.1** *Working fluids: general background*	节 **2.9** 工作流体 节 **2.9.1** 工作流体：基础知识	**SOUS-CHAPITRE 2.9** *Fluides actifs* **SOUS-CHAPITRE 2.9.1** *Généralités sur les fluides actifs*
2.9.1-**1**	azeotrope *azeotropic mixture*	共沸混合物	azéotrope *mélange azéotropique* ◐
2.9.1-**2**	azeotropic mixture *azeotrope*	共沸混合物	azéotrope *mélange azéotropique* ◐
2.9.1-**3**	blend	混合制冷剂	mélange *assemblage* ◐
2.9.1-**4**	compound	化合物	composé
2.9.1-**5**	compression-suction method	压缩-吸入法	méthode (de récupération) par aspiration-compression

ENGLISH	汉语	FRANÇAIS	
coolant *cooling medium* *refrigerating medium*	冷却剂	agent de refroidissement	2.9.1-**6**
cooling medium *coolant* *refrigerating medium*	冷却剂	agent de refroidissement	2.9.1-**7**
cyclic compound	环状化合物	composé cyclique	2.9.1-**8**
fluid quality (x)	流体组分含量比 （x）	concentration massique *titre massique* ◐	2.9.1-**9**
fractionation	分馏	fractionnement	2.9.1-**10**
glide *temperature glide*	（温度）滑移 温度滑移	glissement (de température)	2.9.1-**11**
isomer	（同分、同质）异构体	isomère	2.9.1-**12**
liquid	载冷剂	liquide	2.9.1-**13**
lower flammability limit	可燃性下限	limite inférieure d'inflammabilité	2.9.1-**14**
multi-phase fluid	多相流体	fluide polyphasique	2.9.1-**15**
natural refrigerant	天然工质	fluide (frigorigène) naturel	2.9.1-**16**
near azeotropic	近共沸混合制冷剂	quasi azéotropique	2.9.1-**17**
non-azeotropic mixture *zeotrope* *zeotropic mixture*	非共沸混合物	zéotrope *mélange non azéotropique* ◐ *mélange zéotropique* ◐	2.9.1-**18**
primary fluid *primary refrigerant*	制冷剂 源流体	(fluide) frigorigène primaire	2.9.1-**19**
primary refrigerant *primary fluid*	制冷剂 源流体	(fluide) frigorigène primaire	2.9.1-**20**
reclaim (to)	再生	régénérer	2.9.1-**21**
recover (to)	回收	récupérer	2.9.1-**22**
recovery equipment *refrigerant recovery unit*	制冷剂回收装置	système de récupération de frigorigène	2.9.1-**23**
recycle (to)	循环再利用	recycler	2.9.1-**24**
refrigerant	制冷剂	(fluide) frigorigène	2.9.1-**25**
refrigerant recovery unit *recovery equipment*	制冷剂回收装置	système de récupération de frigorigène	2.9.1-**26**
refrigerating medium *coolant* *cooling medium*	冷却剂	agent de refroidissement	2.9.1-**27**
saturated compound	饱和化合物	composé saturé	2.9.1-**28**
secondary coolant ◐ *secondary refrigerant* *secondary fluid*	载冷剂	fluide secondaire *(fluide) frigoporteur*	2.9.1-**29**
secondary fluid *secondary refrigerant* *secondary coolant* ◐	载冷剂	fluide secondaire *(fluide) frigoporteur*	2.9.1-**30**
secondary refrigerant *secondary fluid* *secondary coolant* ◐	载冷剂	fluide secondaire *(fluide) frigoporteur*	2.9.1-**31**
single-phase fluid	单相流体	fluide monophasique	2.9.1-**32**

	ENGLISH	汉语	FRANÇAIS
2.9.1-**33**	temperature glide *glide*	（温度）滑移 温度滑移	glissement (de température)
2.9.1-**34**	toxicity	毒性	toxicité
2.9.1-**35**	two-phase fluid	两相流体	fluide diphasique
2.9.1-**36**	zeotrope *non-azeotropic mixture* *zeotropic mixture*	非共沸混合物	zéotrope *mélange non azéotropique* ◐ *mélange zéotropique* ◐
2.9.1-**37**	zeotropic mixture *non-azeotropic mixture* *zeotrope*	非共沸混合物	zéotrope *mélange non azéotropique* ◐ *mélange zéotropique* ◐

	SECTION 2.9.2 *Refrigerants* **SECTION 2.9.2.1** *HFCs*	节 **2.9.2** 制冷剂 节 **2.9.2.1** 氢氟烃制冷剂	**SOUS-CHAPITRE 2.9.2** *Fluides frigorigènes* **SOUS-CHAPITRE 2.9.2.1** *HFC*
2.9.2.1-**1**	HFC *hydrofluorocarbon*	氢氟烃	hydrofluorocarbure *HFC*
2.9.2.1-**2**	hydrofluorocarbon *HFC*	氢氟烃	hydrofluorocarbure *HFC*
2.9.2.1-**3**	R125	R125（五氟乙烷）	R125
2.9.2.1-**4**	R134a	R134a（四氟乙烷）	R134a
2.9.2.1-**5**	R14	R14（四氟化碳）	R14
2.9.2.1-**6**	R143a	R143a（三氟乙烷）	R143a
2.9.2.1-**7**	R152a	R152a（二氟乙烷）	R152a
2.9.2.1-**8**	R218	R218（八氟丙烷）	R218
2.9.2.1-**9**	R23	R23（三氟甲烷）	R23
2.9.2.1-**10**	R32	R32（二氟甲烷）	R32
2.9.2.1-**11**	RC318	RC318（八氟环丁烷）	RC318

	SECTION 2.9.2.2 *HCFCs*	节 **2.9.2.2** 氢氯氟烃制冷剂	**SOUS-CHAPITRE 2.9.2.2** *HCFC*
2.9.2.2-**1**	HCFC *hydrochlorofluorocarbon*	氢氯氟烃	hydrochlorofluorocarbure *HCFC*
2.9.2.2-**2**	hydrochlorofluorocarbon *HCFC*	氢氯氟烃	hydrochlorofluorocarbure *HCFC*
2.9.2.2-**3**	R123	R123（三氟二氯乙烷）	R123
2.9.2.2-**4**	R124	R124（四氟一氯乙烷）	R124
2.9.2.2-**5**	R141b	R141b（一氟二氯乙烷）	R141b
2.9.2.2-**6**	R142b	R142b（二氟一氯乙烷）	R142b
2.9.2.2-**7**	R21	R21（一氟二氯甲烷）	R21
2.9.2.2-**8**	R22	R22（二氟一氯甲烷）	R22

ENGLISH	汉语	FRANÇAIS	
SECTION 2.9.2.3 *CFCs*	节 **2.9.2.3** 氯氟烃制冷剂	**SOUS-CHAPITRE 2.9.2.3** *CFC*	
CFC *chorofluorocarbon*	氯氟烃	chlorofluorocarbure *CFC*	2.9.2.3-**1**
chorofluorocarbon *CFC*	氯氟烃	chlorofluorocarbure *CFC*	2.9.2.3-**2**
R11	R11（一氟三氯甲烷）	R11	2.9.2.3-**3**
R12	R12（二氟二氯甲烷）	R12	2.9.2.3-**4**
R13	R13（三氟一氯甲烷）	R13	2.9.2.3-**5**
R13B1	R13B1（三氟一溴甲烷）	R13B1	2.9.2.3-**6**
R113	R113（三氟三氯乙烷）	R113	2.9.2.3-**7**
R114	R114（四氟二氯乙烷）	R114	2.9.2.3-**8**
R115	R115（五氟一氯乙烷）	R115	2.9.2.3-**9**

SECTION 2.9.2.4 *Mixtures*	节 **2.9.2.4** 混合物制冷剂	**SOUS-CHAPITRE 2.9.2.4** *Mélanges*	
R401A	R401A（非共沸混合物）	R401A	2.9.2.4-**1**
R401B	R401B（非共沸混合物）	R401B	2.9.2.4-**2**
R401C	R401C（非共沸混合物）	R401C	2.9.2.4-**3**
R402A	R402A（非共沸混合物）	R402A	2.9.2.4-**4**
R402B	R402B（非共沸混合物）	R402B	2.9.2.4-**5**
R403A	R403A（非共沸混合物）	R403A	2.9.2.4-**6**
R403B	R403B（非共沸混合物）	R403B	2.9.2.4-**7**
R404A	R404A（非共沸混合物）	R404A	2.9.2.4-**8**
R405A	R405A（非共沸混合物）	R405A	2.9.2.4-**9**
R406A	R406A（非共沸混合物）	R406A	2.9.2.4-**10**
R407A	R407A（非共沸混合物）	R407A	2.9.2.4-**11**
R407B	R407B（非共沸混合物）	R407B	2.9.2.4-**12**
R407C	R407C（非共沸混合物）	R407C	2.9.2.4-**13**
R407D	R407D（非共沸混合物）	R407D	2.9.2.4-**14**
R407E	R407E（非共沸混合物）	R407E	2.9.2.4-**15**
R408A	R408A（非共沸混合物）	R408A	2.9.2.4-**16**
R409A	R409A（非共沸混合物）	R409A	2.9.2.4-**17**
R409B	R409B（非共沸混合物）	R409B	2.9.2.4-**18**
R410A	R410A（非共沸混合物）	R410A	2.9.2.4-**19**
R410B	R410B（非共沸混合物）	R410B	2.9.2.4-**20**
R411A	R411A（非共沸混合物）	R411A	2.9.2.4-**21**
R411B	R411B（非共沸混合物）	R411B	2.9.2.4-**22**
R412A	R412A（非共沸混合物）	R412A	2.9.2.4-**23**

	ENGLISH	汉语	FRANÇAIS
2.9.2.4-**24**	R413A	R413A（非共沸混合物）	R413A
2.9.2.4-**25**	R500	R500（共沸混合物）	R500
2.9.2.4-**26**	R502	R502（共沸混合物）	R502
2.9.2.4-**27**	R503	R503（共沸混合物）	R503
2.9.2.4-**28**	R507A	R507（共沸混合物）	R507A
2.9.2.4-**29**	R508A	R508A（共沸混合物）	R508A
2.9.2.4-**30**	R508B	R508B（共沸混合物）	R508B
2.9.2.4-**31**	R509A	R509A（共沸混合物）	R509A

	SECTION 2.9.2.5 *Natural and other refrigerants*	节 **2.9.2.5** 天然制冷剂与其它制冷剂	**SOUS-CHAPITRE 2.9.2.5** *Frigorigènes naturels et autres frigorigènes*
2.9.2.5-**1**	ammonia *R717*	氨 R717（氨）	ammoniac *R717*
2.9.2.5-**2**	ammonium hydroxide	氢氧化铵	hydroxyde d'ammonium *ammoniaque*
2.9.2.5-**3**	butane *R600*	丁烷 R600（丁烷）	butane *R600*
2.9.2.5-**4**	carbon dioxide *R744*	二氧化碳 R744（二氧化碳）	dioxyde de carbone *R744*
2.9.2.5-**5**	ethane *R170*	乙烷 R170（乙烷）	éthane *R170*
2.9.2.5-**6**	ethyl chloride	氯乙烷	chlorure d'éthyle
2.9.2.5-**7**	ethylene *R1150*	乙烯 R1150（乙烯）	éthylène *R1150*
2.9.2.5-**8**	fluorinated hydrocarbon refrigerant *fluorocarbon refrigerant*	氟代烃制冷剂	frigorigène fluorocarboné *frigorigène hydrocarbure fluoré*
2.9.2.5-**9**	fluorocarbon refrigerant *fluorinated hydrocarbon refrigerant*	氟代烃制冷剂	frigorigène fluorocarboné *frigorigène hydrocarbure fluoré*
2.9.2.5-**10**	halocarbon	卤代烃	hydrocarbure halogéné *halocarbure*
2.9.2.5-**11**	HBFC *hydrobromofluorocarbon*	氢溴氟烃	hydrobromofluorocarbure *HBFC*
2.9.2.5-**12**	HC *hydrocarbon*	烃 碳氢化合物	hydrocarbure *HC*
2.9.2.5-**13**	HFE *hydrofluoroether*	氢氟醚	hydrofluoroéther *HFE*
2.9.2.5-**14**	hydrobromofluorocarbon *HBFC*	氢溴氟烃	hydrobromofluorocarbure *HBFC*
2.9.2.5-**15**	hydrocarbon *HC*	烃 碳氢化合物	hydrocarbure *HC*
2.9.2.5-**16**	hydrofluoroether *HFE*	氢氟醚	hydrofluoroéther *HFE*
2.9.2.5-**17**	isobutane *R600a*	异丁烷 R600a（异丁烷）	isobutane *R600a*
2.9.2.5-**18**	methane *R50*	甲烷 R50（甲烷）	méthane *R50*

ENGLISH	汉语	FRANÇAIS	
methyl chloride	氯甲烷 甲基氯	chlorure de méthyle	2.9.2.5-**19**
methylene chloride	二氯甲烷	chlorure de méthylène	2.9.2.5-**20**
perfluorocarbon *PFC*	全氟代烃	perfluorocarbure *PFC*	2.9.2.5-**21**
PFC *perfluorocarbon*	全氟代烃	perfluorocarbure *PFC*	2.9.2.5-**22**
propane *R290*	丙烷 R290（丙烷）	propane *R290*	2.9.2.5-**23**
propylene *R1270*	丙烯 R1270（丙烯）	propylène *R1270*	2.9.2.5-**24**
R1150 *ethylene*	乙烯 R1150（乙烯）	éthylène *R1150*	2.9.2.5-**25**
R1270 *propylene*	丙烯 R1270（丙烯）	propylène *R1270*	2.9.2.5-**26**
R170 *ethane*	乙烷 R170（乙烷）	éthane *R170*	2.9.2.5-**27**
R290 *propane*	丙烷 R290（丙烷）	propane *R290*	2.9.2.5-**28**
R50 *methane*	甲烷 R50（甲烷）	méthane *R50*	2.9.2.5-**29**
R600 *butane*	丁烷 R600（丁烷）	butane *R600*	2.9.2.5-**30**
R600a *isobutane*	异丁烷 R600a（异丁烷）	isobutane *R600a*	2.9.2.5-**31**
R717 *ammonia*	氨 R717（氨）	ammoniac *R717*	2.9.2.5-**32**
R718 *water*	水 R718（水）	eau *R718*	2.9.2.5-**33**
R744 *carbon dioxide*	二氧化碳 R744（二氧化碳）	dioxyde de carbone *R744*	2.9.2.5-**34**
R764 *surphur dioxide*	二氧化硫 R764（二氧化硫）	dioxyde de soufre *R764*	2.9.2.5-**35**
sulphur dioxide *R764*	二氧化硫 R764（二氧化硫）	dioxyde de soufre *R764*	2.9.2.5-**36**
water *R718*	水 R718（水）	eau *R718*	2.9.2.5-**37**

SECTION 2.9.3 *Secondary refrigerants* **SECTION 2.9.3.1** *Brine*	节 **2.9.3** 载冷剂 节 **2.9.3.1** 盐水	**SOUS-CHAPITRE 2.9.3** *Fluides frigoporteurs* **SOUS-CHAPITRE 2.9.3.1** *Saumure*	
agitator	搅拌器	agitateur	2.9.3.1-**1**
antifreeze agent	防冻剂	antigel	2.9.3.1-**2**
aqueous solution	水溶液	solution aqueuse	2.9.3.1-**3**
brine *non-freeze liquid* *non-freezing solution*	盐水 不冻液	saumure *liquide à bas point de congélation* *liquide "incongelable"* ◐	2.9.3.1-**4**

	ENGLISH	汉语	FRANÇAIS
2.9.3.1-**5**	brine balance tank *brine expansion tank* *brine head tank* ◐	盐水膨胀箱	vase d'expansion (de saumure) *bac d'expansion de saumure*
2.9.3.1-**6**	brine cooler	盐水冷却器	refroidisseur de saumure
2.9.3.1-**7**	brine drum	盐水筒	tube accumulateur de saumure
2.9.3.1-**8**	brine expansion tank *brine balance tank* *brine head tank* ◐	盐水膨胀箱	vase d'expansion (de saumure) *bac d'expansion de saumure*
2.9.3.1-**9**	brine header	盐水集管	collecteur de saumure
2.9.3.1-**10**	brine head tank ◐ *brine balance tank* *brine expansion tank*	盐水膨胀箱	vase d'expansion (de saumure) *bac d'expansion de saumure*
2.9.3.1-**11**	brine mixing tank	盐水混合箱	bac de mélange de saumure
2.9.3.1-**12**	brine pump	盐水泵	pompe à saumure
2.9.3.1-**13**	brine return tank	盐水回流箱	réservoir à retour de saumure
2.9.3.1-**14**	brine sparge	盐水喷淋	distributeur de saumure à aspersion
2.9.3.1-**15**	brine spray	盐水喷雾	pulvérisation de saumure
2.9.3.1-**16**	brine tank	盐水箱	bac à saumure
2.9.3.1-**17**	closed-brine system	闭式盐水箱	système fermé à saumure
2.9.3.1-**18**	cooling mixture	冷却混合物	mélange refroidisseur
2.9.3.1-**19**	corrosion inhibitor	缓蚀剂	inhibiteur de corrosion
2.9.3.1-**20**	densimeter *hydrometer* ◐ *salinometer* ◐ *twaddle gauge* ○	比重计 密度计	densimètre *aréomètre* ○ *pèse-saumure* ○
2.9.3.1-**21**	eutectic mixture	共晶混合物	mélange eutectique
2.9.3.1-**22**	freezant	冻结剂	médium congélateur
2.9.3.1-**23**	freezing mixture	冻结混合物	mélange congélateur
2.9.3.1-**24**	heat-transfer fluid *heat-transfer medium*	传热流体 传热介质	fluide caloporteur
2.9.3.1-**25**	heat-transfer medium *heat-transfer fluid*	传热流体 传热介质	fluide caloporteur
2.9.3.1-**26**	hydrometer ◐ *densimeter* *salinometer* ◐ *twaddle gauge* ○	比重计 密度计	densimètre *aréomètre* ○ *pèse-saumure* ○
2.9.3.1-**27**	non-freeze liquid *brine* *non-freezing solution*	盐水 不冻液	saumure *liquide à bas point de congélation* *liquide "incongelable"* ◐
2.9.3.1-**28**	non-freezing solution *brine* *non-freeze liquid*	盐水 不冻液	saumure *liquide à bas point de congélation* *liquide "incongelable"* ◐
2.9.3.1-**29**	open-brine system	开式盐水系统	système ouvert à saumure
2.9.3.1-**30**	salinometer ◐ *densimeter* *hydrometer* ◐ *twaddle gauge* ○	比重计 密度计	densimètre *aréomètre* ○ *pèse-saumure* ○
2.9.3.1-**31**	secondary loop	二次回流	circuit frigoporteur *circuit secondaire* ◐

ENGLISH	汉语	FRANÇAIS	
twaddle gauge ○ *densimeter* *hydrometer* ◐ *salinometer* ◐	比重计 密度计	densimètre *aréomètre* ○ *pèse-saumure* ○	2.9.3.1-**32**

SECTION 2.9.3.2 *Chilled and iced water*	节 **2.9.3.2** 冷水与冰水	**SOUS-CHAPITRE 2.9.3.2** *Eau réfrigérée et eau glacée*	
chilled water	冷水	eau réfrigérée *eau glacée* ◐	2.9.3.2-**1**
ice bank tank *ice build-up tank*	蓄冰槽	bac à accumulation de glace	2.9.3.2-**2**
ice build-up tank *ice bank tank*	蓄冰槽	bac à accumulation de glace	2.9.3.2-**3**
ice slurry *pumpable ice*	冰浆 可用泵输送的冰	coulis de glace *(coulis de) glace "pompable"*	2.9.3.2-**4**
ice water	融冰水	eau de fusion de la glace	2.9.3.2-**5**
iced water	冰水	eau glacée	2.9.3.2-**6**
iced water tank	冰水槽	bac à eau glacée	2.9.3.2-**7**
pumpable ice *ice slurry*	冰浆 可用泵输送的冰	coulis de glace *(coulis de) glace "pompable"*	2.9.3.2-**8**

SECTION 2.9.3.3 *Other secondary refrigerants*	节 **2.9.3.3** 其它载冷剂	**SOUS-CHAPITRE 2.9.3.3** *Autres frigoporteurs*	
aqua ammonia	氨水	liqueur ammoniacale *alcali*	2.9.3.3-**1**
calcium chloride	氯化钙	chlorure de calcium	2.9.3.3-**2**
carbon dioxide snow *CO_2 snow*	二氧化碳雪 固体二氧化碳	neige carbonique	2.9.3.3-**3**
CO_2 snow *carbon dioxide snow*	二氧化碳雪 固体二氧化碳	neige carbonique	2.9.3.3-**4**
cryohydrate	低共融冰盐结晶 冰盐	cryohydrate	2.9.3.3-**5**
dry ice	干冰 固体二氧化碳	glace sèche *glace carbonique*	2.9.3.3-**6**
ethanol *ethyl alcohol* ◐	乙醇 酒精	éthanol *alcool éthylique*	2.9.3.3-**7**
ethyl alcohol ◐ *ethanol*	乙醇 酒精	éthanol *alcool éthylique*	2.9.3.3-**8**
ethylene glycol	乙二醇 甘醇	éthylène glycol	2.9.3.3-**9**
eutectic ice *frozen eutectic solution*	共晶冰	glace eutectique *solution eutectique congelée*	2.9.3.3-**10**
frozen eutectic solution *eutectic ice*	共晶冰	glace eutectique *solution eutectique congelée*	2.9.3.3-**11**
glycol water	乙二醇水	eau glycolée	2.9.3.3-**12**

	ENGLISH	汉语	FRANÇAIS
2.9.3.3-**13**	magnesium chloride	氯化镁	chlorure de magnésium
2.9.3.3-**14**	methanol *methyl alcohol*	甲醇	méthanol *alcool méthylique*
2.9.3.3-**15**	methyl alcohol *methanol*	甲醇	méthanol *alcool méthylique*
2.9.3.3-**16**	potassium acetate	乙酸钾	acétate de potassium
2.9.3.3-**17**	potassium carbonate	碳酸钾	carbonate de potassium
2.9.3.3-**18**	propylene glycol	丙烯钾	propylène glycol
2.9.3.3-**19**	sodium chloride	氯化钠	chlorure de sodium

	SECTION 2.9.4 *Other working fluids*	节 **2.9.4** 其它工作流体	**SOUS-CHAPITRE 2.9.4** *Autres fluides actifs*
2.9.4-**1**	activated carbon	活性碳	charbon actif
2.9.4-**2**	ammonia-water	氨-水	couple eau-ammoniac
2.9.4-**3**	lithium bromide	溴化锂	bromure de lithium
2.9.4-**4**	lithium bromide-water	溴化锂-水	couple eau-bromure de lithium
2.9.4-**5**	silica gel	硅胶	gel de silice
2.9.4-**6**	zeolite	沸石	zéolite *zéolithe*

章 3. 制冷设备

◐ 容许（或默许）术语

○ 过时术语

ENGLISH	汉语	FRANÇAIS	
SECTION 3.1 *Refrigerating equipment: general background*	节 **3.1** 制冷设备：基础知识	**SOUS-CHAPITRE 3.1** *Généralités sur les installations frigorifiques*	
cascade refrigerating system	复叠式制冷系统	installation frigorifique en cascade *système cascade*	**3.1-1**
central refrigerating plant	集中制冷系统	centrale frigorifique *installation centrale de froid* ◐	**3.1-2**
commercial refrigerating plant	商用制冷装置	installation frigorifique commerciale	**3.1-3**
engine room ◐ *machinery room* *machine room* ◐	机器间 机房	salle des machines	**3.1-4**
factory-assembled system ◐ *packaged unit* *self-contained system*	成套机组	installation frigorifique autonome *installation frigorifique monobloc* *système frigorifique préassemblé*	**3.1-5**
human-occupied space	人居空间	enceinte occupée par des personnes	**3.1-6**
industrial refrigerating plant	工业制冷装置	installation frigorifique industrielle	**3.1-7**
machine room ◐ *machinery room* *engine room* ◐	机器间 机房	salle des machines	**3.1-8**
machinery room *engine room* ◐ *machine room* ◐	机器间 机房	salle des machines	**3.1-9**
multistage refrigerating plant	多级制冷系统	installation frigorifique étagée *système multi-étagé*	**3.1-10**
packaged unit *self-contained system* *factory-assembled system* ◐	成套机组	installation frigorifique autonome *installation frigorifique monobloc* *système frigorifique préassemblé*	**3.1-11**
refrigerating equipment	制冷设备	composant frigorifique	**3.1-12**
refrigerating installation ◐ *refrigerating plant*	制冷装置	installation frigorifique	**3.1-13**
refrigerating loop	制冷回路	circuit frigorifique	**3.1-14**
refrigerating plant *refrigerating installation* ◐	制冷装置	installation frigorifique	**3.1-15**
self-contained system *packaged unit* *factory-assembled system* ◐	成套机组	installation frigorifique autonome *installation frigorifique monobloc* *système frigorifique préassemblé*	**3.1-16**

SECTION 3.2 *Construction*	节 **3.2** 建造	**SOUS-CHAPITRE 3.2** *Construction*	
access door *access hatch* ○	便门	portillon de service	**3.2-1**
access hatch ○ *access door*	便门	portillon de service	**3.2-2**
air curtain	空气幕	rideau d'air	**3.2-3**
beam *girder* ○	横梁	poutre	**3.2-4**
cargo battens ◐ *wall dunnage* ◐	护墙格栅	vaigrage à claire-voie (marine)	**3.2-5**

	ENGLISH	汉语	FRANÇAIS
3.2-6	cavity brick construction ○	空心砖建筑	maçonnerie de corps creux *parpaings creux*
3.2-7	cavity wall ○	空心墙	mur à double paroi *mur à vide d'air* ◐
3.2-8	crawl space *guard space* ◐	架空层	vide sanitaire
3.2-9	curtain wall	轻质墙	mur rideau
3.2-10	door gasket	门密封垫	joint de porte
3.2-11	dunnage	垫木	lattes d'arrimage *lattis d'arrimage*
3.2-12	flexible door	弹簧门	porte souple
3.2-13	floor dunnage ◐ *floor rack* *floor grating* ◐	垫仓板	caillebotis
3.2-14	floor grating ◐ *floor rack* *floor dunnage* ◐	垫仓板	caillebotis
3.2-15	floor rack *floor dunnage* ◐ *floor grating* ◐	垫仓板	caillebotis
3.2-16	flush fitting door *infitting door* ◐	嵌式门	porte encastrée
3.2-17	frame wall ◐	框架墙	mur à ossature
3.2-18	framework	框架	charpente
3.2-19	frost heave ◐	地坪冻鼓	soulèvement du sol par congélation
3.2-20	girder ○ *beam*	横梁	poutre
3.2-21	guard space ◐ *crawl space*	架空层	vide sanitaire
3.2-22	hard covering *wearing surface (of a floor)* ◐	耐磨层	chape d'usure
3.2-23	heater cable ◐ *heater strip* *heater tape* ○	加热条	bande chauffante *câble chauffant* *cordon chauffant*
3.2-24	heater mat	加热层	nappe chauffante *réseau antigel*
3.2-25	heater strip *heater cable* ◐ *heater tape* ○	加热条	bande chauffante *câble chauffant* *cordon chauffant*
3.2-26	heater tape ○ *heater strip* *heater cable* ◐	加热条	bande chauffante *câble chauffant* *cordon chauffant*
3.2-27	hung ceiling ◐ *suspended ceiling*	吊顶	plafond suspendu
3.2-28	infitting door ◐ *flush fitting door*	嵌式门	porte encastrée
3.2-29	inspection window *porthole* ◐	观察窗	hublot
3.2-30	insulated door	隔热门	porte isolante *porte isolée*

ENGLISH	汉语	FRANÇAIS	
insulated openings	隔热洞口	menuiseries isolantes *menuiseries isothermes* ○	3.2-**31**
insulated web (of a girder) ○	隔热梁腹	âme isolante (de poutre)	3.2-**32**
masonry wall	砖石墙	mur en maçonnerie	3.2-**33**
mushroom floor	无梁楼板	plancher champignon	3.2-**34**
overhead rail	吊轨	rail aérien	3.2-**35**
overlap door	外贴门	porte en surépaisseur *porte en applique* ◐	3.2-**36**
panelling	护壁板	bardage *parement*	3.2-**37**
peep-hole	窥视孔	judas	3.2-**38**
plug door	嵌式检查门	tampon de visite	3.2-**39**
porthole ◐ *inspection window*	观察窗	hublot	3.2-**40**
powered door	电动门	porte commandée	3.2-**41**
protective coating	保护层	revêtement de protection	3.2-**42**
purlin ○	檩条	panne	3.2-**43**
rafter ◐	椽条	chevron	3.2-**44**
sliding door	推拉门	porte coulissante	3.2-**45**
soil freezing	土壤冻结	congélation du sol	3.2-**46**
suspended ceiling *hung ceiling* ◐	吊顶	plafond suspendu	3.2-**47**
swinging door	转动门	porte battante *porte va-et-vient*	3.2-**48**
truss (of a roof) ○	屋架	ferme	3.2-**49**
underfloor ventilation	地下通风	ventilation sous plancher *plancher soufflant*	3.2-**50**
ventilated crawl space	通风敷设通道	vide sanitaire ventilé	3.2-**51**
wall dunnage ◐ *cargo battens* ◐	护墙格栅	vaigrage à claire-voie (marine)	3.2-**52**
wall rail	墙 栏杆	lisse	3.2-**53**
wearing surface (of a floor) ◐ *hard covering*	耐磨层	chape d'usure	3.2-**54**

SECTION 3.3 *Thermal insulation* **SECTION 3.3.1** *Thermal insulation: general background*	节 **3.3** 隔热 节 **3.3.1** 隔热：基础知识	**SOUS-CHAPITRE 3.3** *Isolation thermique* **SOUS-CHAPITRE 3.3.1** *Généralités sur l'isolation thermique*	
backing insulation	带（保护）衬板的隔热	isolation protégée	3.3.1-**1**
blanket-type insulant	毡状隔热材料	isolant en matelas souple	3.3.1-**2**
block-type insulant	块状隔热材料	isolant en blocs	3.3.1-**3**

	ENGLISH	汉语	FRANÇAIS
3.3.1-**4**	board-type insulant ◐ *slab insulant* ◐	板状隔热材料	isolant en panneau *isolant en plaque*
3.3.1-**5**	composite insulation	组合型隔热	isolation composite
3.3.1-**6**	conventional insulation	常规隔热	isolation traditionnelle
3.3.1-**7**	evacuated insulation ◐ *vacuum insulation*	真空绝热	isolation sous vide
3.3.1-**8**	fill insulation *loose-fill-type insulant* ○	填充式隔热	isolant de bourrage *isolant en vrac* *isolant de remplissage*
3.3.1-**9**	foil insulant	箔隔热材料	isolant en feuilles
3.3.1-**10**	heat bridge *heat channel* ◐	热桥	pont thermique
3.3.1-**11**	heat channel ◐ *heat bridge*	热桥	pont thermique
3.3.1-**12**	heat leakage	热渗漏	fuite thermique *déperdition de chaleur* ◐
3.3.1-**13**	insulate (to)	隔热（动词）	calorifuger *isoler*
3.3.1-**14**	insulated	隔热的	isolé
3.3.1-**15**	insulating jacket	隔热夹套	enveloppe isolante
3.3.1-**16**	jacket	夹套	double paroi
3.3.1-**17**	loose-fill-type insulant ○ *fill insulation*	填充式隔热	isolant de bourrage *isolant en vrac* *isolant de remplissage*
3.3.1-**18**	mat-type insulant	垫状隔热材料	isolant en matelas
3.3.1-**19**	powdered insulant	粉末隔热材料	isolant en poudre
3.3.1-**20**	pre-formed insulation	预成型隔热	isolant préformé
3.3.1-**21**	slab insulant ◐ *board-type insulant* ◐	板状隔热材料	isolant en panneau *isolant en plaque*
3.3.1-**22**	superinsulation	超绝热	superisolation
3.3.1-**23**	thermal inertia	热惰性	inertie thermique
3.3.1-**24**	(thermal) insulation	隔热	calorifugeage *isolation (thermique)* ◐
3.3.1-**25**	thermal insulation	隔热	isolation thermique
3.3.1-**26**	thermal insulation composite system	组合型隔热	système d'isolation thermique composite
3.3.1-**27**	unbound insulation	非粘合型隔热	isolant non encollé
3.3.1-**28**	vacuum insulation *evacuated insulation* ◐	真空绝热	isolation sous vide
3.3.1-**29**	wall losses	墙壁热损失	pertes (thermiques) par les parois

	SECTION 3.3.2 *Insulating materials*	节 **3.3.2** 隔热材料	**SOUS-CHAPITRE 3.3.2** *matériaux isolants*
3.3.2-**1**	aluminium foil	铝箔	feuille d'aluminium
3.3.2-**2**	blanket *mat*	毯子 毡子 垫子	feutre

ENGLISH	汉语	FRANÇAIS	
blowing wool	吹（送）纤维	isolant particulaire insufflable	3.3.2-**3**
bonding	粘合	ensimage	3.3.2-**4**
calcium silicate	硅酸钙	silicate de calcium	3.3.2-**5**
carbon fibre	碳纤维	fibre de carbone	3.3.2-**6**
cellular concrete ◐ *foam concrete* *foamed concrete*	泡沫混凝土	béton cellulaire	3.3.2-**7**
cellular glass *glass foam*	泡沫玻璃	verre cellulaire *verre mousse*	3.3.2-**8**
cellular insulant ◐	多孔隔热材料	isolant cellulaire	3.3.2-**9**
cellular material	多孔材料	matériau alvéolaire	3.3.2-**10**
cellular plastic *plastic foam* ◐	泡沫塑料	mousse plastique *plastique alvéolaire* *plastique cellulaire*	3.3.2-**11**
cellular rubber ◐	泡沫橡胶	caoutchouc cellulaire *caoutchouc mousse*	3.3.2-**12**
cellulose insulation	纤维隔热	isolant cellulosique	3.3.2-**13**
ceramic fibre	陶瓷纤维	fibre céramique	3.3.2-**14**
closed-cell foamed plastic	闭孔泡沫塑料	plastique cellulaire à cellules fermées	3.3.2-**15**
composite insulation product	组合隔热制品	produit isolant composite	3.3.2-**16**
composite panel	组合板	panneau composite	3.3.2-**17**
cork	软木	liège	3.3.2-**18**
cork board	软木板	panneau de liège	3.3.2-**19**
diatomaceous earth ○	硅藻土	terre à diatomées	3.3.2-**20**
elastomer	弹性体	élastomère	3.3.2-**21**
evacuated powder	真空粉末	poudre sous vide	3.3.2-**22**
expanded (cellular) plastic	膨胀（多孔）塑料	plastique (cellulaire) expansé	3.3.2-**23**
expanded clay	膨胀粘土	argile expansée	3.3.2-**24**
expanded cork	膨胀软木	liège expansé	3.3.2-**25**
expanded perlite board	膨胀珍珠岩板	panneau de perlite expansée	3.3.2-**26**
expanded polystyrene	膨胀聚苯乙烯	polystyrène expansé	3.3.2-**27**
expanded polyvinyl chloride	膨胀聚氯乙烯	mousse de PVC	3.3.2-**28**
extruded (cellular) plastic	挤压（多孔）塑料	plastique (cellulaire) extrudé	3.3.2-**29**
extruded polystyrene foam	挤压多孔聚苯乙烯	mousse de polystyrène extrudé	3.3.2-**30**
fibrous insulant	纤维隔热材料	isolant fibreux	3.3.2-**31**
fibrous insulation	纤维隔热	isolation fibreuse	3.3.2-**32**
flexible elastomeric foam	柔性泡沫合成橡胶	mousse souple élastomère	3.3.2-**33**
foam concrete *foamed concrete* *cellular concrete* ◐	泡沫混凝土	béton cellulaire	3.3.2-**34**
foamed concrete *foam concrete* *cellular concrete* ◐	泡沫混凝土	béton cellulaire	3.3.2-**35**
gas space	气隙	lame d'air	3.3.2-**36**

	ENGLISH	汉语	FRANÇAIS
3.3.2-**37**	glass fibre *glass wool*	玻璃棉 玻璃纤维	fibre de verre *laine de verre* ◐
3.3.2-**38**	glass foam *cellular glass*	泡沫玻璃	verre cellulaire *verre mousse*
3.3.2-**39**	glass wool *glass fibre*	玻璃棉 玻璃纤维	fibre de verre *laine de verre* ◐
3.3.2-**40**	granulated cork	粒状软木	granulés de liège *liège granulé*
3.3.2-**41**	granulated wool	松散纤维	laine en flocons ou en nodules
3.3.2-**42**	graphite fibre	石墨纤维	fibre de graphite
3.3.2-**43**	in situ thermal insulation product	现场制隔热产品	produit d'isolation thermique in situ
3.3.2-**44**	insulant ◐ *insulating material*	隔热材料	isolant (subst.) *matériau isolant*
3.3.2-**45**	insulating ◐	隔热	isolant (adj.)
3.3.2-**46**	insulating castable refractory	可浇铸的隔热耐火材料	béton réfractaire isolant
3.3.2-**47**	insulating concrete *lightweight concrete*	轻质混凝土	béton allégé *béton de granulats légers* *béton isolant*
3.3.2-**48**	insulating material *insulant* ◐	隔热材料	isolant (subst.) *matériau isolant*
3.3.2-**49**	insulating plaster	隔热涂层	plâtre isolant
3.3.2-**50**	insulating rope	隔热绳	bourrelet isolant
3.3.2-**51**	lightweight aggregate	轻质填充料	granulat léger
3.3.2-**52**	lightweight concrete *insulating concrete*	轻质混凝土	béton allégé *béton de granulats légers* *béton isolant*
3.3.2-**53**	magnesia	氧化镁	magnésie
3.3.2-**54**	man-made mineral fibre	人造矿渣纤维	fibre minérale manufacturée
3.3.2-**55**	mat *blanket*	毯子 毡子 垫子	feutre
3.3.2-**56**	microporous insulation	微孔（材料）隔热	isolant microporeux *aérogel de silice*
3.3.2-**57**	mineral fibre	矿渣纤维	fibre minérale
3.3.2-**58**	mineral wool *slag wool* ◐	矿渣棉	laine de laitier *laine minérale*
3.3.2-**59**	multicellular metal foil	多孔金属箔	feuille métallique à structure cellulaire
3.3.2-**60**	multilayer insulant	多层隔热材料	isolant multicouche
3.3.2-**61**	opacified silica-aerogel	不透热硅石一气凝胶	aérosilicagel opacifié
3.3.2-**62**	open-cell foamed plastic	开孔泡沫塑料	plastique cellulaire à cellules ouvertes
3.3.2-**63**	perlite	珍珠岩	perlite
3.3.2-**64**	perlite plaster	珍珠岩涂层	plâtre de perlite
3.3.2-**65**	phenolic foam	酚醛泡沫塑料	mousse phénolique
3.3.2-**66**	pipe insulation	管道隔热	isolation de tuyauterie
3.3.2-**67**	plastic foam ◐ *cellular plastic*	泡沫塑料	mousse plastique *plastique alvéolaire* *plastique cellulaire*

ENGLISH	汉语	FRANÇAIS	
polyethylene foam	泡沫聚乙烯	mousse de polyéthylène	3.3.2-**68**
polyisocyanurate foam	聚异氰酸酯泡沫	mousse polyisocyanurate	3.3.2-**69**
polyurethane foam	聚氨酯泡沫	mousse rigide de polyuréthane	3.3.2-**70**
pouring wool	充填纤维	laine à déverser	3.3.2-**71**
prefabricated panel	予制（隔热）板	panneau préfabriqué	3.3.2-**72**
reflective insulant	反射隔热材料	isolant réfléchissant	3.3.2-**73**
reflective insulation	反射隔热	isolation réfléchissante	3.3.2-**74**
rock wool	石棉	laine de roche	3.3.2-**75**
sandwich panel	夹心（隔热）板	panneau sandwich	3.3.2-**76**
sandwich panel insulation	夹心板隔热	isolation en panneaux	3.3.2-**77**
slag wool ◐ *mineral wool*	矿渣棉	laine de laitier *laine minérale*	3.3.2-**78**
slotted slab	带槽沟的（隔热）板	panneau rainuré	3.3.2-**79**
thermal insulation material	隔热材料	matériau d'isolation thermique	3.3.2-**80**
thermal insulation product	隔热制品	produit d'isolation thermique	3.3.2-**81**
urea formaldehyde foam	脲醛泡沫塑料	mousse urée-formaldéhyde	3.3.2-**82**
vacuum insulation jacket	真空隔热夹套	enveloppe isolante sous vide	3.3.2-**83**
vacuum reflective insulation	真空反射隔热	isolation réfléchissante sous vide	3.3.2-**84**
vermiculite	蛭石	vermiculite	3.3.2-**85**
wood wool	木纤维	laine de bois	3.3.2-**86**
wood wool board	木纤维板	panneau en laine de bois	3.3.2-**87**

SECTION 3.3.3 *Installation of insulants*	节 **3.3.3** 隔热材料的装配	**SOUS-CHAPITRE 3.3.3** *Installation des matériaux isolants*	
binder	粘合剂	liant	3.3.3-**1**
block	块状（隔热材料）	bloc isolant	3.3.3-**2**
blown insulation	松散隔热	isolant par soufflage	3.3.3-**3**
board slab	板状（隔热材料） 板块	panneau isolant	3.3.3-**4**
breaker strip ◐	隔热杆	entretoise isolante *barrette de maintien* ◐ *rupteur de conduction* ◐	3.3.3-**5**
cladding	（金属）面板	revêtement	3.3.3-**6**
coating	涂层	enduit de finition	3.3.3-**7**
curved board	弧形板	panneau incurvé	3.3.3-**8**
elbow	弯头	coude	3.3.3-**9**
embedded insulation	埋置隔热	isolation en fond de coffrage	3.3.3-**10**
facing	面层	parement	3.3.3-**11**
finishing cement	水泥抹面	enduit de finition *enduit de finition (ciment)* *enduit hydraulique de finition*	3.3.3-**12**

	ENGLISH	汉语	FRANÇAIS
3.3.3-13	foamed in-place insulation *foamed in-situ insulation* ◐	现场发泡隔热	isolation expansée in situ
3.3.3-14	foamed in-situ insulation ◐ *foamed in-place insulation*	现场发泡隔热	isolation expansée in situ
3.3.3-15	insulated suspending tiebar	隔热悬挂杆	suspente isolante
3.3.3-16	insulating brick	隔热砖	brique isolante
3.3.3-17	insulating cement	隔热粘结剂	enduit isolant *ciment isolant*
3.3.3-18	insulating joint	隔热节	joint de retrait isolant
3.3.3-19	insulating mastic	隔热玛王帝脂	mastic isolant
3.3.3-20	insulation cover	隔热套	coquille isolante
3.3.3-21	insulation finish	隔热保护层	enduit pour isolation
3.3.3-22	laminate	层压板	produit feuilleté *produit laminé* ◐
3.3.3-23	mattress	（隔热）垫	matelas isolant
3.3.3-24	mitred joint	斜角接合	onglet
3.3.3-25	moulding	模制	bandelette isolante
3.3.3-26	pipe section	隔热管段	fourreau isolant
3.3.3-27	pneumatic application	气力输送	application pneumatique
3.3.3-28	poured application	人工填充	application par déversement *application par remplissage*
3.3.3-29	radiation shield	辐射护罩	écran antirayonnement
3.3.3-30	roll	缠绕	rouleau
3.3.3-31	self-supporting insulation	自承重隔热	isolation autoportante
3.3.3-32	slab *board*	板状（隔热材料） 板块	panneau isolant
3.3.3-33	sprayed insulation ◐ *sprayed-on insulation* ◐	喷涂隔热	isolation par projection
3.3.3-34	sprayed-on insulation ◐ *sprayed insulation* ◐	喷涂隔热	isolation par projection
3.3.3-35	structural insulation	结构隔热	isolation porteuse
3.3.3-36	tube insulation	管道用软性隔热材料	fourreau isolant *manchon isolant*
3.3.3-37	wire mat	铁丝网	nappe grillagée

	SECTION 3.3.4 *Gas and vapour seals*	节 **3.3.4** 气体与蒸气密封	**SOUS-CHAPITRE 3.3.4** *Etanchéité à la vapeur et aux gaz*
3.3.4-1	accumulation test	累积试验	contrôle par accumulation
3.3.4-2	airtight	气密的	étanche à l'air
3.3.4-3	asphalted paper	沥青纸	papier bitumé
3.3.4-4	bituminous felt	沥青毡	feutre bitumineux

ENGLISH	汉语	FRANÇAIS	
breather plug *vent* ◐	呼吸孔	évent	3.3.4-**5**
calibration leak	渗漏校验	fuite calibrée	3.3.4-**6**
capillary leak	毛细渗漏	fuite capillaire	3.3.4-**7**
coating barrier ◐	隔气涂层	enduit d'étanchéité	3.3.4-**8**
damp proofing ◐ *moisture proofing*	防潮	imperméabilisation	3.3.4-**9**
gas-proof *gas-tight*	气密的	étanche aux gaz	3.3.4-**10**
gas seal	气密层	écran d'étanchéité aux gaz	3.3.4-**11**
gas-tight *gas-proof*	气密的	étanche aux gaz	3.3.4-**12**
laminated paper	层合纸	papier stratifié	3.3.4-**13**
leak-free ◐ *leaktight*	密封的 无渗漏的	étanche *imperméable*	3.3.4-**14**
leakage rate	渗漏率	flux de fuite	3.3.4-**15**
leaktight *leak-free* ◐	密封的 无渗漏的	étanche *imperméable*	3.3.4-**16**
membrane barrier	薄膜密封层	feuille d'étanchéité *membrane d'étanchéité*	3.3.4-**17**
moisture barrier *vapour barrier* *vapour seal* *water vapour barrier*	防潮层 隔汽层	barrière anti-vapeur *écran d'étanchéité à la vapeur (d'eau)* *écran pare-vapeur*	3.3.4-**18**
moisture proofing *damp proofing* ◐	防潮	imperméabilisation	3.3.4-**19**
permeability	渗透性	perméabilité	3.3.4-**20**
permeability coefficient	渗透系数	coefficient de perméabilité	3.3.4-**21**
permeance	渗透率	perméance	3.3.4-**22**
plastic film	塑料薄膜	pellicule plastique	3.3.4-**23**
sealing mastic *sealing putty* *sealing stopper* ◐	密封玛王帝脂	mastic d'étanchéité	3.3.4-**24**
sealing putty *sealing mastic* *sealing stopper* ◐	密封玛王帝脂	mastic d'étanchéité	3.3.4-**25**
sealing stopper ◐ *sealing mastic* *sealing putty*	密封玛王帝脂	mastic d'étanchéité	3.3.4-**26**
structural barrier	结构密封层	étanchéité structurale	3.3.4-**27**
vapour barrier *moisture barrier* *vapour seal* *water vapour barrier*	防潮层 隔汽层	barrière anti-vapeur *écran d'étanchéité à la vapeur (d'eau)* *écran pare-vapeur*	3.3.4-**28**
vapour migration *vapour transmission* *vapour transfer* ◐	蒸汽迁移	migration de la vapeur (d'eau) *passage de la vapeur (d'eau)* *transfert de la vapeur (d'eau)*	3.3.4-**29**
vapour permeability	蒸气渗透性	perméabilité à la vapeur (d'eau)	3.3.4-**30**
vapour-proof ◐ *vapour-tight*	气密的	étanche à la vapeur (d'eau) *imperméable à la vapeur (d'eau)*	3.3.4-**31**

	ENGLISH	汉语	FRANÇAIS
3.3.4-**32**	vapour seal *moisture barrier* *vapour barrier* *water vapour barrier*	防潮层 隔汽层	barrière anti-vapeur *écran d'étanchéité à la vapeur (d'eau)* *écran pare-vapeur*
3.3.4-**33**	vapour-tight *vapour-proof* ◐	气密的	étanche à la vapeur (d'eau) *imperméable à la vapeur (d'eau)*
3.3.4-**34**	vapour transfer ◐ *vapour migration* *vapour transmission*	蒸汽迁移	migration de la vapeur (d'eau) *passage de la vapeur (d'eau)* *transfert de la vapeur (d'eau)*
3.3.4-**35**	vapour transmission *vapour migration* *vapour transfer* ◐	蒸汽迁移	migration de la vapeur (d'eau) *passage de la vapeur (d'eau)* *transfert de la vapeur (d'eau)*
3.3.4-**36**	vent ◐ *breather plug*	呼吸孔	évent
3.3.4-**37**	water vapour barrier *moisture barrier* *vapour barrier* *vapour seal*	防潮层 隔汽层	barrière anti-vapeur *écran d'étanchéité à la vapeur (d'eau)* *écran pare-vapeur*
3.3.4-**38**	water vapour retarder	水汽隔离层	pare-vapeur

	SECTION 3.4 *Sound insulation*	节 **3.4** 隔声	**SOUS-CHAPITRE 3.4** *Isolation phonique*
3.4-**1**	acoustic insulation *sound insulation* ◐	隔声	isolation acoustique *isolation phonique*
3.4-**2**	damping (of vibration)	（振动）衰减	amortissement (de vibration)
3.4-**3**	muffler (USA) ◐ *silencer* *sound attenuator* *noise damper* ◐ *sound absorber* ◐	消声器	silencieux *insonorisateur*
3.4-**4**	noise criteria curves	噪声判据曲线	courbes de bruit
3.4-**5**	noise damper ◐ *silencer* *sound attenuator* *muffler (USA)* ◐ *sound absorber* ◐	消声器	silencieux *insonorisateur*
3.4-**6**	noise reduction *sound attenuation* *sound damping* ◐ *sound deadening* ◐	声衰减 消声	insonorisation *amortissement du son* *atténuation du bruit*
3.4-**7**	silencer *sound attenuator* *muffler (USA)* ◐ *noise damper* ◐ *sound absorber* ◐	消声器	silencieux *insonorisateur*
3.4-**8**	sound absorber ◐ *silencer* *sound attenuator* *muffler (USA)* ◐ *noise damper* ◐	消声器	silencieux *insonorisateur*
3.4-**9**	sound attenuation *noise reduction* *sound damping* ◐ *sound deadening* ◐	声衰减 消声	insonorisation *amortissement du son* *atténuation du bruit*

ENGLISH	汉语	FRANÇAIS	
sound attenuator *silencer* *muffler (USA)* ◐ *noise damper* ◐ *sound absorber* ◐	消声器	silencieux *insonorisateur*	3.4-**10**
sound damping ◐ *noise reduction* *sound attenuation* *sound deadening* ◐	声衰减 消声	insonorisation *amortissement du son* *atténuation du bruit*	3.4-**11**
sound deadening ◐ *noise reduction* *sound attenuation* *sound damping* ◐	声衰减 消声	insonorisation *amortissement du son* *atténuation du bruit*	3.4-**12**
sound insulation ◐ *acoustic insulation*	隔声	isolation acoustique *isolation phonique*	3.4-**13**
sound level	声级	niveau sonore	3.4-**14**
sound power level (L_W)	声功率级 (L_W)	niveau de puissance acoustique (L_W)	3.4-**15**
sound pressure level	声压级	niveau de pression acoustique	3.4-**16**
sound trap	声阱	piège à sons	3.4-**17**

SECTION 3.5 *Operation* **SECTION 3.5.1** *Running of refrigerating equipment*	节 **3.5** 运行 节 **3.5.1** 制冷设备的运行	**SOUS-CHAPITRE 3.5** *Exploitation* **SOUS-CHAPITRE 3.5.1** *Fonctionnement des installations frigorifiques*	
actuate (to) ◐	起动（动词）	actionner ◐	3.5.1-**1**
automatic starting system	自动起动系统	appareillage de démarrage automatique *circuit de démarrage automatique* ◐	3.5.1-**2**
balance pressure	平衡压力	pression d'équilibre	3.5.1-**3**
belt drive	皮带传动	entraînement par courroie	3.5.1-**4**
breakdown (of a machine)	（机器）故障	panne (de machine)	3.5.1-**5**
burn-out (of a motor)	（电机）烧毁	grillage (d'un moteur)	3.5.1-**6**
calculation temperature	计算温度	température de calcul	3.5.1-**7**
capacity control *capacity modulation*	能量调节	régulation de la puissance (frigorifique) *variation de la puissance (d'un compresseur)*	3.5.1-**8**
capacity controller *capacity regulator* ◐	能量调节器	dispositif de variation de puissance *régulateur de puissance*	3.5.1-**9**
capacity modulation *capacity control*	能量调节	régulation de la puissance (frigorifique) *variation de la puissance (d'un compresseur)*	3.5.1-**10**
capacity reducer ○	能量调节器	réducteur de puissance	3.5.1-**11**
capacity regulator ◐ *capacity controller*	能量调节器	dispositif de variation de puissance *régulateur de puissance*	3.5.1-**12**
cavitation	气蚀	cavitation	3.5.1-**13**
charging	充注	charge (d'un circuit)	3.5.1-**14**
commissioning	试运转	mise en (état de) fonctionnement	3.5.1-**15**
control system	控制系统	système de régulation	3.5.1-**16**

	ENGLISH	汉语	FRANÇAIS
3.5.1-**17**	corrosion	腐蚀	corrosion
3.5.1-**18**	degassing ◐ *gas purging* *purging* *gas-off* ◐	气体排除 放气	dégazage *purge*
3.5.1-**19**	design pressure *maximum allowable pressure* *maximum declared pressure* *maximum working pressure (MWP)*	设计（工作）压力 设计压力 最高允许压力 最高公称压力 最高工作压力	pression maximale de service
3.5.1-**20**	design temperature	设计温度	température de conception spécifiée
3.5.1-**21**	direct drive	直接传动	(entraînement par) accouplement direct
3.5.1-**22**	drive	传动	entraînement
3.5.1-**23**	drive-through a gearbox	齿轮传动	entraînement par boîte de vitesse *entraînement par une pignonnerie*
3.5.1-**24**	flash gas	闪发气体	vapeur "instantanée" *vaporisat* ◐
3.5.1-**25**	flash vaporization *instantaneous vaporization* ◐	闪发	vaporisation instantanée
3.5.1-**26**	flush (to)	吹除（动词）	faire une chasse
3.5.1-**27**	flywheel	飞轮	volant
3.5.1-**28**	freeze-up	冻堵	bouchage par congélation
3.5.1-**29**	gas bottle (1) ◐ *refrigerant cylinder*	制冷剂瓶	bouteille de frigorigène (de livraison)
3.5.1-**30**	gas bottle (2) ◐ *service cylinder* ◐	维修用制冷剂瓶	bouteille de frigorigène (de dépanneur)
3.5.1-**31**	gas lock *vapour lock*	气囊	bouchon de vapeur
3.5.1-**32**	gas-off ◐ *gas purging* *purging* *degassing* ◐	气体排除 放气	dégazage *purge*
3.5.1-**33**	gas purging *purging* *degassing* ◐ *gas-off* ◐	气体排除 放气	dégazage *purge*
3.5.1-**34**	gas shortage ◐ *lack of refrigerant* *undercharge*	制冷剂不足	manque de fluide
3.5.1-**35**	gauge (1)	表	jauge
3.5.1-**36**	gauge (2)	量规	calibre
3.5.1-**37**	hammering *liquid hammer* *pipe hammer* ◐	水锤现象	coup de bélier
3.5.1-**38**	hammering (of an expansion valve) ◐ *needle hammer*	阀针跳动（膨胀阀的）	martèlement d'un obturateur
3.5.1-**39**	holding charge *service charge* ◐	保护充注	charge d'attente
3.5.1-**40**	hunting (of a valve)	（阀）振荡	pompage (d'un régleur)
3.5.1-**41**	initial charge	首次充注	charge initiale *première charge*

ENGLISH	汉语	FRANÇAIS	
instantaneous vaporization ◐ *flash vaporization*	闪发	vaporisation instantanée	3.5.1-**42**
inverter drive	变频传动	entraînement des moteurs par variateur de fréquence	3.5.1-**43**
lack of refrigerant *undercharge* *gas shortage* ◐	制冷剂不足	manque de fluide	3.5.1-**44**
leak	渗漏	fuite	3.5.1-**45**
liquid hammer *hammering* *pipe hammer* ◐	水锤现象	coup de bélier	3.5.1-**46**
mains (supply) (UK) *power (supply) (USA)* ◐	电源	alimentation électrique	3.5.1-**47**
manual starting system	人工起动系统	démarrage manuel	3.5.1-**48**
maximum allowable pressure *design pressure* *maximum declared pressure* *maximum working pressure (MWP)*	设计（工作）压力 设计压力 最高允许压力 最高公称压力 最高工作压力	pression maximale de service	3.5.1-**49**
maximum declared pressure *design pressure* *maximum allowable pressure* *maximum working pressure (MWP)*	设计（工作）压力 设计压力 最高允许压力 最高公称压力 最高工作压力	pression maximale de service	3.5.1-**50**
maximum working pressure (MWP) *design pressure* *maximum allowable pressure* *maximum declared pressure*	设计（工作）压力 设计压力 最高允许压力 最高公称压力 最高工作压力	pression maximale de service	3.5.1-**51**
monitoring	监控	surveillance (technique)	3.5.1-**52**
multiple-speed governor controller	多速控制器	régulateur à vitesse multiple	3.5.1-**53**
needle hammer *hammering (of an expansion valve)* ◐	阀针跳动（膨胀阀的）	martèlement d'un obturateur	3.5.1-**54**
no-load starting ◐ *unloaded start* ◐	卸载起动	démarrage à vide	3.5.1-**55**
non-condensable gas	不凝性气体	gaz non condensable *incondensable*	3.5.1-**56**
operating conditions	运行工况	régime de fonctionnement *régime de marche*	3.5.1-**57**
operating pressure *working pressure*	工作压力	pression en service *pression de service*	3.5.1-**58**
operating temperature	工作温度	température en service	3.5.1-**59**
overcharge	过量充注	excès de charge *excès de fluide* ◐	3.5.1-**60**
overfeeding	供液过量	suralimentation	3.5.1-**61**
overflow	溢液管 溢流器	trop-plein	3.5.1-**62**
overheating	异常升温	échauffement (anormal) *surchauffe (d'un équipement)*	3.5.1-**63**
pipe hammer ◐ *hammering* *liquid hammer*	水锤现象	coup de bélier	3.5.1-**64**

	ENGLISH	汉语	FRANÇAIS
3.5.1-**65**	power (supply) (USA) ◐ *mains (supply) (UK)*	电源	alimentation électrique
3.5.1-**66**	power take-off	动力分配器	prise de force
3.5.1-**67**	pressure equalizing	压力平衡	équilibrage des pressions
3.5.1-**68**	pull-down test	降温试验	essai de mise en régime
3.5.1-**69**	pump down (of refrigerant)	（制冷剂的）抽空	évacuation (du frigorigène)
3.5.1-**70**	purging *gas purging* *degassing* ◐ *gas-off* ◐	气体排除 放气	dégazage *purge*
3.5.1-**71**	rapid cycling	快速循环	fonctionnement en cycles rapides
3.5.1-**72**	refrigerant charge	制冷剂充注量	charge de fluide (frigorigène)
3.5.1-**73**	refrigerant cylinder *gas bottle (1)* ◐	制冷剂瓶	bouteille de frigorigène (de livraison)
3.5.1-**74**	refrigeration running test	制冷性能试验	essai frigorifique *essai de refroidissement*
3.5.1-**75**	running cycle	运行周期	période
3.5.1-**76**	service charge ◐ *holding charge*	保护充注	charge d'attente
3.5.1-**77**	service cylinder ◐ *gas bottle (2)* ◐	维修用制冷剂瓶	bouteille de frigorigène (de dépanneur)
3.5.1-**78**	short-cycling	短循环	fonctionnement en courts cycles *courts cycles*
3.5.1-**79**	single-speed governor controller	单速控制器	régulateur à vitesse unique
3.5.1-**80**	slugging	液击	coup de liquide
3.5.1-**81**	starting air valve	起动空气阀	soupape de démarrage
3.5.1-**82**	starting interlock	起动联锁装置	sécurité de démarrage
3.5.1-**83**	starting system	起动系统	système de démarrage
3.5.1-**84**	starving (of an evaporator)	（蒸发器）缺液	sous-alimentation (d'un évaporateur)
3.5.1-**85**	strength-test pressure	强度测试压力 试验压力	pression de l'essai de résistance *pression d'épreuve* ◐
3.5.1-**86**	test bed ○ *test bench* *test stand* *test rig* ◐	试验台	banc d'essai
3.5.1-**87**	test bench *test stand* *test rig* ◐ *test bed* ○	试验台	banc d'essai
3.5.1-**88**	test rig ◐ *test bench* *test stand* *test bed* ○	试验台	banc d'essai
3.5.1-**89**	test stand *test bench* *test rig* ◐ *test bed* ○	试验台	banc d'essai
3.5.1-**90**	test temperature	试验温度	température lors de l'essai de résistance
3.5.1-**91**	trouble shooting	故障分析	diagnostic

ENGLISH	汉语	FRANÇAIS	
undercharge *lack of refrigerant* *gas shortage* ◐	制冷剂不足	manque de fluide	3.5.1-**92**
unloaded start ◐ *no-load starting* ◐	卸载起动	démarrage à vide	3.5.1-**93**
unloader	卸载机构	bipasse de démarrage *dispositif de délestage*	3.5.1-**94**
v-belt	三角皮带	courroie trapézoïdale	3.5.1-**95**
vacuum test	真空试验	essai sous vide	3.5.1-**96**
valve bounce ◐ *valve flutter*	阀片跳动	battement (d'un clapet)	3.5.1-**97**
valve flutter *valve bounce* ◐	阀片跳动	battement (d'un clapet)	3.5.1-**98**
vapour lock *gas lock*	气囊	bouchon de vapeur	3.5.1-**99**
variable-speed governor controller	变速控制器	régulateur à vitesse variable *régulateur toute vitesse* ◐	3.5.1-**100**
working pressure *operating pressure*	工作压力	pression en service *pression de service*	3.5.1-**101**

SECTION 3.5.2 *Lubrication*	节 **3.5.2** 润滑	**SOUS-CHAPITRE 3.5.2** *Lubrification*	
autogenous ignition temperature ○ *spontaneous ignition temperature*	自燃温度	température d'auto-inflammation	3.5.2-**1**
breakdown (of an oil)	（油）变质	décomposition (d'une huile) *altération*	3.5.2-**2**
carbonization	碳化	carbonisation	3.5.2-**3**
cloud point	浊点	point de trouble	3.5.2-**4**
dip lubrication	点滴润滑	lubrification par barbotage	3.5.2-**5**
drop point	滴点	point de goutte	3.5.2-**6**
fire point	燃点	point d'inflammation	3.5.2-**7**
flash point	闪点	point d'éclair	3.5.2-**8**
flock point	絮凝点	point de floculation	3.5.2-**9**
foaming	起泡	moussage	3.5.2-**10**
forced-feed oiling ○ *forced lubrication* *mechanical lubrication* ◐ *pump lubrication* ◐	强制润滑	lubrification sous pression *lubrification forcée*	3.5.2-**11**
forced lubrication *mechanical lubrication* ◐ *pump lubrication* ◐ *forced-feed oiling* ○	强制润滑	lubrification sous pression *lubrification forcée*	3.5.2-**12**
lubrication	润滑	lubrification *graissage*	3.5.2-**13**
mechanical lubrication ◐ *forced lubrication* *pump lubrication* ◐ *forced-feed oiling* ○	强制润滑	lubrification sous pression *lubrification forcée*	3.5.2-**14**

	ENGLISH	汉语	FRANÇAIS
3.5.2-**15**	oil charge	充油量	charge d'huile
3.5.2-**16**	oil content	（制冷剂中的）油含量	teneur en huile
3.5.2-**17**	oil cooler	油冷却器	refroidisseur d'huile
3.5.2-**18**	oil distributor	油道	circuit d'huile (d'une machine)
3.5.2-**19**	oil gauge ○ *oil sight glass*	油位指示器	indicateur de niveau d'huile *niveau d'huile* ◐
3.5.2-**20**	oil level	油位	niveau d'huile
3.5.2-**21**	oil level indicator	油位指示器	indicateur de niveau d'huile *niveau d'huile* ◐
3.5.2-**22**	oil pressure	油压	pression d'huile
3.5.2-**23**	oil pressure gauge	油压表	manomètre de pression d'huile
3.5.2-**24**	oil pump	油泵	pompe à huile
3.5.2-**25**	oil removal	除油	déshuilage
3.5.2-**26**	oil return	回油	retour d'huile
3.5.2-**27**	oil sight glass *oil gauge* ○	油位指示器	indicateur de niveau d'huile *niveau d'huile* ◐
3.5.2-**28**	pour point	流动点	point d'écoulement
3.5.2-**29**	pump lubrication ◐ *forced lubrication* *mechanical lubrication* ◐ *forced-feed oiling* ○	强制润滑	lubrification sous pression *lubrification forcée*
3.5.2-**30**	sludge	沉淀物	boues
3.5.2-**31**	softening point	软化点	point de ramollissement
3.5.2-**32**	splash lubrication	飞溅润滑	lubrification par projection
3.5.2-**33**	spontaneous ignition temperature *autogenous ignition temperature* ○	自燃温度	température d'auto-inflammation
3.5.2-**34**	viscosity index	粘度指数	indice de viscosité
3.5.2-**35**	wax content	含腊量	teneur en paraffine

	SECTION 3.5.3 *Defrosting*	节 **3.5.3** 除霜	**SOUS-CHAPITRE 3.5.3** *Dégivrage*
3.5.3-**1**	automatic defrosting	自动除霜	dégivrage automatique
3.5.3-**2**	defrost pan (USA) *drip tray*	除霜水盘	bac de dégivrage *cuvette de dégivrage*
3.5.3-**3**	defrost time	除霜时间	durée de dégivrage
3.5.3-**4**	defrost water removal	去除除霜水	évacuation de l'eau de dégivrage
3.5.3-**5**	defrosting	除霜	dégivrage
3.5.3-**6**	defrosting cycle	除霜周期	cycle de dégivrage
3.5.3-**7**	drip tray *defrost pan (USA)*	除霜水盘	bac de dégivrage *cuvette de dégivrage*
3.5.3-**8**	electric defrosting	电热除霜	dégivrage électrique
3.5.3-**9**	external defrosting	外能除霜	dégivrage par l'extérieur
3.5.3-**10**	frost	霜	givre

ENGLISH	汉语	FRANÇAIS	
frost back ◐	回霜	givrage à l'aspiration (d'un compresseur) ◐	3.5.3-**11**
frost deposit ○	积霜	dépôt de givre ○	3.5.3-**12**
frost formation	结霜	givrage	3.5.3-**13**
frosted ◐	有霜的	givré ◐	3.5.3-**14**
hot-gas defrosting	热气除霜	dégivrage par gaz chauds	3.5.3-**15**
internal defrosting	内能除霜	dégivrage par l'intérieur	3.5.3-**16**
manual defrosting	人工除霜	dégivrage manuel	3.5.3-**17**
off-cycle defrosting	中止循环除霜	dégivrage naturel cyclique *dégivrage à chaque cycle*	3.5.3-**18**
pressure defrosting	压力除霜	dégivrage commandé par la perte de pression sur l'air	3.5.3-**19**
reverse-cycle defrosting	逆循环除霜	dégivrage par cycle inversé *dégivrage par inversion de cycle*	3.5.3-**20**
semi-automatic defrosting	半自动除霜	dégivrage semi-automatique	3.5.3-**21**
thermobank defrost	热箱除霜	thermo-accumulateur pour dégivrage	3.5.3-**22**
time defrosting	定时除霜	dégivrage chronocommandé *dégivrage (commandé) par chronorelais*	3.5.3-**23**
time-temperature defrosting	定时定温除霜	dégivrage à commande combinée par le temps et la température *dégivrage chrono-thermique*	3.5.3-**24**
water defrosting	水除霜	dégivrage par (aspersion d') eau *dégivrage par (ruissellement d') eau*	3.5.3-**25**

章 4. 冷却方法

◐ 容许（或默许）术语

○ 过时术语

ENGLISH	汉语	FRANÇAIS	
SECTION 4.1 *Cooling methods: general background*	节 **4.1** 冷却方法：基础知识	**SOUS-CHAPITRE 4.1** *Généralités sur les méthodes de refroidissement*	
air blast cooling	强制通风冷却	refroidissement par air soufflé *refroidissement par soufflage d'air*	4.1-**1**
contact cooling	接触冷却	refroidissement par contact	4.1-**2**
cooling bath ○	冷浸液	bain de refroidissement	4.1-**3**
cooling method *cooling process* ◐	冷却方法	méthode de refroidissement *mode de refroidissement*	4.1-**4**
cooling process ◐ *cooling method*	冷却方法	méthode de refroidissement *mode de refroidissement*	4.1-**5**
direct expansion refrigeration *direct refrigerating system*	直接膨胀制冷系统 直接式制冷系统	système de refroidissement direct *refroidissement par détente directe*	4.1-**6**
direct refrigerating system *direct expansion refrigeration*	直接膨胀制冷系统 直接式制冷系统	système de refroidissement direct *refroidissement par détente directe*	4.1-**7**
equilibration temperature	平衡温度	température d'équilibre *température d'équilibre (thermique)*	4.1-**8**
evaporative cooling	蒸发冷却	refroidissement évaporatif *refroidissement par évaporation d'eau*	4.1-**9**
forced-draught cooling ○	强制通风冷却	refroidissement par convection forcée *refroidissement par ventilation forcée*	4.1-**10**
immersion cooling	沉浸冷却	refroidissement par immersion	4.1-**11**
indirect refrigerating system	间接式制冷系统 间接制冷系统 间接冷却系统	système de refroidissement indirect *système secondaire de refroidissement* *système à fluide secondaire* ◐	4.1-**12**
natural-convection cooling	自然对流冷却	refroidissement par convection naturelle	4.1-**13**
partial-recovery refrigeration	部分回收式制冷系统	refroidissement à frigorigène partiellement récupéré	4.1-**14**
recessed fitting	嵌入式安装	montage encastré	4.1-**15**
spray cooling	喷淋冷却	refroidissement par aspersion	4.1-**16**
surface cooling	表面冷却	refroidissement sur surface froide	4.1-**17**
total-loss refrigeration	不回收式（开式）制冷系统	refroidissement à frigorigène perdu	4.1-**18**
transpiration cooling	蒸发冷却	refroidissement par transpiration	4.1-**19**
vacuum chilling ◐ *vacuum cooling*	真空冷却	réfrigération par le vide *refroidissement par le vide*	4.1-**20**
vacuum cooling *vacuum chilling* ◐	真空冷却	réfrigération par le vide *refroidissement par le vide*	4.1-**21**

ENGLISH	汉语	FRANÇAIS	
SECTION 4.2 *Chilling*	节 **4.2** 冷却	**SOUS-CHAPITRE 4.2** *Réfrigération*	
chill (to)	冷却（动词）	réfrigérer	4.2-**1**
chilling	冷却	réfrigération	4.2-**2**
cooling down	降温	opération de refroidissement *tombée en froid* ◐	4.2-**3**
cooling rate	降温速率	vitesse de refroidissement	4.2-**4**

	ENGLISH	汉语	FRANÇAIS
4.2-5	flood-type hydrocooling	满液式水冷却	réfrigération par immersion dans l'eau
4.2-6	fogging	成雾	brumisation *nébulisation* ◐
4.2-7	half-cooling time	半冷却时间	temps de demi-refroidissement
4.2-8	hydrocooling	水冷却	réfrigération par eau glacée
4.2-9	impingement cooling *jet cooling* ◐	喷气冷却	refroidissement par jet d'air froid
4.2-10	jacket cooling	夹套式冷却	réfrigération par double paroi *réfrigération par enveloppe froide* ◐
4.2-11	jet cooling ◐ *impingement cooling*	喷气冷却	refroidissement par jet d'air froid
4.2-12	liquid-chilling package	液体冷却机组	groupe refroidisseur de liquide
4.2-13	precooler	预冷器	prérefroidisseur
4.2-14	precooling	预冷	préréfrigération
4.2-15	pressure cooling	压力冷却	réfrigération par pression d'air
4.2-16	quick chilling ◐ *rapid chilling*	快速冷却	réfrigération rapide
4.2-17	rapid chilling *quick chilling* ◐	快速冷却	réfrigération rapide
4.2-18	refrigerator	冷柜	réfrigérateur
4.2-19	spray-type hydrocooling	喷淋式水冷却	réfrigération par aspersion d'eau glacée
4.2-20	superchilling	过冷	superréfrigération

	SECTION 4.3 *Freezing*	节 **4.3** 冻结	**SOUS-CHAPITRE 4.3** *Congélation*
4.3-1	air-blast freezer	强制通风冻结设备	congélateur à air soufflé
4.3-2	air-blast freezing	强制通风冻结	congélation par air soufflé
4.3-3	batch freezer	间歇式冻结设备	congélateur discontinu
4.3-4	belt freezer	皮带式冻结器	congélateur à convoyeur
4.3-5	belt freezing	传送带冻结	congélation sur convoyeur
4.3-6	bulk freezing	散装冻结	congélation en vrac
4.3-7	contact freezer	接触式冻结设备	congélateur par contact
4.3-8	contact freezing	接触冻结	congélation par contact
4.3-9	continuous freezer	连续式冻结器	congélateur continu
4.3-10	crust freezing *shell freezing* ◐	表层冻结	croûtage *congélation superficielle* ◐
4.3-11	cryogenic freezer	深冷冻结设备	congélateur (à froid) cryogénique
4.3-12	deep freezing ◐ *quick freezing*	快速冻结	surgélation *congélation rapide* ◐
4.3-13	deep-freezing plant ◐ *quick-freezing plant*	快速冻结装置	installation de surgélation *installation de congélation rapide* ◐
4.3-14	deep-frozen food ◐ *quick-frozen food*	速冻食品 快速冻结食品	denrées surgelées *aliments surgelés*

ENGLISH	汉语	FRANÇAIS	
dehydro-freezing	脱水冻结	déshydratation-congélation	4.3-**15**
dielectric thawing	电介质解冻	décongélation diélectrique	4.3-**16**
double-contact freezer	双面接触式冻结设备	congélateur par double contact	4.3-**17**
drip	汁液流失	exsudat	4.3-**18**
dual freezing system	双冻结系统	congélateur à froid mixte	4.3-**19**
effective freezing time	有效冻结时间	durée effective de congélation *temps effectif de congélation*	4.3-**20**
fluidized bed	流态床	lit fluidisé	4.3-**21**
fluidized-bed freezer	流态床冻结设备	congélateur à lit fluidisé	4.3-**22**
fluidized-bed freezing *fluidized freezing* ○	流态床冻结	congélation en lit fluidisé	4.3-**23**
fluidized freezing ○ *fluidized-bed freezing*	流态床冻结	congélation en lit fluidisé	4.3-**24**
freeze (to) (foods)	（食品）冻结（动词）	congeler (des aliments)	4.3-**25**
freezer *freezer unit* ◐	冻结设备	congélateur	4.3-**26**
freezer capacity	冻结设备容量	charge d'un congélateur	4.3-**27**
freezer unit ◐ *freezer*	冻结设备	congélateur	4.3-**28**
freezing capacity	冻结能力	capacité de congélation	4.3-**29**
freezing method *freezing process* ○	冻结方法	méthode de congélation *mode de congélation*	4.3-**30**
freezing plant	冻结装置	installation de congélation	4.3-**31**
freezing plateau	冻结稳定	palier de congélation	4.3-**32**
freezing process ○ *freezing method*	冻结方法	méthode de congélation *mode de congélation*	4.3-**33**
freezing rate *speed of freezing* ◐	冻结速度	vitesse de congélation	4.3-**34**
freezing room	冻结间	chambre de congélation	4.3-**35**
freezing section	冻结工段	atelier de congélation	4.3-**36**
freezing time	冻结时间	durée de congélation	4.3-**37**
freezing tunnel *tunnel freezer* ◐	冻结隧道	tunnel de congélation	4.3-**38**
frozen food	冻结食品	aliment congelé *denrée congelée*	4.3-**39**
frozen-food locker ○	带锁冷冻食品柜（冷库中）	casier congélateur *casier pour denrées congelées*	4.3-**40**
glazing	包冰衣	glaçage *givrage*	4.3-**41**
high-frequency thawing	高频解冻	décongélation par haute fréquence	4.3-**42**
ice-slurry freezing	冰浆冻结	congélation dans un coulis de glace	4.3-**43**
immersion freezer	沉浸式冻结装置	congélateur par immersion	4.3-**44**
immersion freezing	沉浸冻结	congélation par immersion	4.3-**45**
impingement freezing *jet freezing* ○	喷气冻结	congélation par jet d'air	4.3-**46**
individual quick freezing (IQF)	单体快速冻结	surgélation individuelle	4.3-**47**

	ENGLISH	汉语	FRANÇAIS
4.3-**48**	initial freezing temperature ◐ *temperature of initial ice formation*	初始冻结温度	température de congélation commençante
4.3-**49**	initial melting temperature	起始融化温度	température de fusion commençante
4.3-**50**	jet freezing ○ *impingement freezing*	喷气冻结	congélation par jet d'air
4.3-**51**	lift off (of particles)	（颗粒）吹失	envol (de particules)
4.3-**52**	loose freezing	松散冻结	congélation en masse divisée
4.3-**53**	mechanical freezer	机械制冷冻结器	congélateur à froid mécanique
4.3-**54**	microwave thawing *ultra-high frequency thawing* ○	微波解冻	décongélation par microondes *décongélation par hyperfréquence* ◐
4.3-**55**	multiplate freezer	多板冻结设备	congélateur à plaques multiples
4.3-**56**	nominal freezing time	公称冻结时间	durée nominale de congélation
4.3-**57**	plate freezer	平板冻结设备	congélateur à plaque
4.3-**58**	quick freezing *deep freezing* ◐	快速冻结	surgélation *congélation rapide* ◐
4.3-**59**	quick-freezing plant *deep-freezing plant* ◐	快速冻结装置	installation de surgélation *installation de congélation rapide* ◐
4.3-**60**	quick-frozen food *deep-frozen food* ◐	速冻食品 快速冻结食品	denrées surgelées *aliments surgelés*
4.3-**61**	rated freezing capacity	额定冻结能力	aptitude de congélation nominale
4.3-**62**	refreezing	再冻	recongélation
4.3-**63**	refrigerated locker ○	带锁冷冻食品柜（冷库中）	case frigorifique *casier frigorifique*
4.3-**64**	scraped-surface freezer	刮式冻结装置	congélateur à surface raclée *congélateur à racleur* ◐
4.3-**65**	shelf freezer	搁架式冻结装置	congélateur à étagères
4.3-**66**	shell freezing ◐ *crust freezing*	表层冻结	croûtage *congélation superficielle* ◐
4.3-**67**	slow freezing	慢速冻结	congélation lente
4.3-**68**	speed of freezing ◐ *freezing rate*	冻结速度	vitesse de congélation
4.3-**69**	spray freezer	喷淋式冻结装置	congélateur par aspersion *congélateur par pulvérisation*
4.3-**70**	spray freezing	喷淋式冻结	congélation par aspersion *congélation par pulvérisation*
4.3-**71**	still-air freezing	在静止的空气中冻结	congélation en air calme
4.3-**72**	temperature of initial ice formation *initial freezing temperature* ◐	初始冻结温度	température de congélation commençante
4.3-**73**	thaw (to)	解冻（动词）	décongeler *dégeler*
4.3-**74**	thawing	解冻	décongélation
4.3-**75**	thawing time	解冻时间	durée de décongélation
4.3-**76**	thermal centre	热中心	centre thermique
4.3-**77**	tunnel freezer ◐ *freezing tunnel*	冻结隧道	tunnel de congélation
4.3-**78**	ultra-high frequency thawing ○ *microwave thawing*	微波解冻	décongélation par microondes *décongélation par hyperfréquence* ◐

ENGLISH	汉语	FRANÇAIS	
SECTION 4.4 *Vacuum application*	节 **4.4** 真空处理	**SOUS-CHAPITRE 4.4** *Traitement sous vide*	
cold trap	冷阱	piège froid	**4.4-1**
condensation (vacuum) pump ○ *diffusion (vacuum) pump*	扩散（真空）泵	pompe (à vide) à condensation *pompe (à vide) à diffusion* ◐	**4.4-2**
diffusion (vacuum) pump *condensation (vacuum) pump* ○	扩散（真空）泵	pompe (à vide) à condensation *pompe (à vide) à diffusion* ◐	**4.4-3**
flash point	闪点	point de vaporisation *point d'évaporation*	**4.4-4**
ion (vacuum) pump	离子（真空）泵	pompe (à vide) ionique	**4.4-5**
liquid mechanical (vacuum) pump	液体机械（真空）泵	pompe (à vide) à anneau liquide	**4.4-6**
sorption (vacuum) pump	吸附（真空）泵	pompe (à vide) à sorption	**4.4-7**
vacuum breaking *vacuum cracking* ◐	真空破坏	cassage du vide	**4.4-8**
vacuum cracking ◐ *vacuum breaking*	真空破坏	cassage du vide	**4.4-9**
vacuum device	真空设备	appareil sous vide	**4.4-10**
vacuum plant ◐ *vacuum system*	真空系统	installation de vide	**4.4-11**
vacuum pump	真空泵	pompe à vide	**4.4-12**
vacuum system *vacuum plant* ◐	真空系统	installation de vide	**4.4-13**
vapour (vacuum) pump	蒸气（真空）泵	pompe (à vide) à jet de vapeur *pompe à vapeur de… (mercure, huile…)*	**4.4-14**

章 5. 储藏 、运输、分配

◐ 容许（或默许）术语

○ 过时术语

ENGLISH	汉语	FRANÇAIS	
SECTION 5.1 *Domestic equipment*	节 **5.1** 家用设备	**SOUS-CHAPITRE 5.1** *Equipements ménagers*	
absorption refrigerator	吸收式冰箱	réfrigérateur à absorption	5.1-**1**
beverage dispenser	自动饮料出售机	distributeur de boissons	5.1-**2**
bottle beverage cooler *bottle cooler* ◐	瓶装饮料冷却器	refroidisseur de bouteilles	5.1-**3**
bottle cooler ◐ *bottle beverage cooler*	瓶装饮料冷却器	refroidisseur de bouteilles	5.1-**4**
built-in refrigerator	嵌入式冰箱	réfrigérateur encastré *réfrigérateur intégré* *réfrigérateur incorporé* ◐	5.1-**5**
cooling device	冷却设备	dispositif de refroidissement	5.1-**6**
domestic freezer *household food freezer* *home freezer* ◐	家用冷冻柜 家用食品冷冻柜	congélateur ménager *congélateur domestique* ◐	5.1-**7**
domestic refrigerator *household refrigerator*	家用冰箱	réfrigérateur ménager *armoire frigorifique ménagère* *réfrigérateur domestique* ◐	5.1-**8**
door racks	门衬架	balconnet	5.1-**9**
drinking-water cooler *fountain (USA)* ◐ *water dispenser* ○	饮用水冷却器	rafraîchisseur d'eau (potable) *fontaine réfrigérée*	5.1-**10**
dual-temperature refrigerator	双温冰箱	réfrigérateur-congélateur	5.1-**11**
enamel (paint)	烤漆	peinture laquée	5.1-**12**
fountain (USA) ◐ *drinking water cooler* *water dispenser* ○	饮用水冷却器	rafraîchisseur d'eau (potable) *fontaine réfrigérée*	5.1-**13**
freezer compartment *low-temperature compartment* ◐	低温格 低温室	compartiment congélateur *compartiment à basse température*	5.1-**14**
freezer total gross volume	总的毛容积	volume brut total d'un congélateur	5.1-**15**
freezer total storage volume	总的贮存容积	volume utile total d'un congélateur	5.1-**16**
gas refrigerator	煤气冰箱	réfrigérateur à gaz	5.1-**17**
home freezer ◐ *domestic freezer* *household food freezer*	家用冷冻柜 家用食品冷冻柜	congélateur ménager *congélateur domestique* ◐	5.1-**18**
household food freezer *domestic freezer* *home freezer* ◐	家用冷冻柜 家用食品冷冻柜	congélateur ménager *congélateur domestique* ◐	5.1-**19**
household frozen-food storage cabinet ("three-star" cabinet)	家用冷冻食品贮藏柜	conservateur conservateur de denrées congelées à usage ménager (conservateur "trois étoiles")	5.1-**20**
household refrigerator *domestic refrigerator*	家用冰箱	réfrigérateur ménager *armoire frigorifique ménagère* *réfrigérateur domestique* ◐	5.1-**21**
ice box ◐ *ice refrigerator*	冰冷冰箱	glacière domestique *réfrigérateur à glace*	5.1-**22**
ice (cube) tray	方块冰盘	tiroir à glace *tiroir à glaçons*	5.1-**23**
ice refrigerator *ice box* ◐	冰冷冰箱	glacière domestique *réfrigérateur à glace*	5.1-**24**

	ENGLISH	汉语	FRANÇAIS
5.1-25	inner cabinet liner *liner (of a refrigerator)* ◐	冷柜内壳	cuve intérieure (d'un réfrigérateur) *cuve interne (d'un réfrigérateur)* ◐
5.1-26	liner (of a refrigerator) ◐ *inner cabinet liner*	冷柜内壳	cuve intérieure (d'un réfrigérateur) *cuve interne (d'un réfrigérateur)* ◐
5.1-27	load-limit volume ◐	装载容积限量	volume intérieur brut *volume de charge maximal*
5.1-28	low-temperature compartment ◐ *freezer compartment*	低温格 低温室	compartiment congélateur *compartiment à basse température*
5.1-29	nursery refrigerator	保育室用冰箱	réfrigérateur de biberons
5.1-30	open refrigerator	敞开式冷柜	présentoir frigorifique
5.1-31	overall dimensions (doors or lids closed)	外型尺寸（关闭门或盖）	dimensions hors-tout (portes ou couvercles fermés) *hors-tout*
5.1-32	overall space required in use (doors or lids open)	使用所需空间（门或盖敞开）	encombrement en service (portes ou couvercles ouverts)
5.1-33	rated energy consumption	额定能耗	consommation d'énergie nominale
5.1-34	rated gross volume	额定毛容积	volume brut nominal
5.1-35	rated storage volume	额定贮藏容积	volume utile nominal
5.1-36	rated total gross volume	额定总的毛容积	volume brut total nominal
5.1-37	rated total storage volume	额定总的贮藏容积	volume utile total nominal
5.1-38	self-contained refrigerator	自携式冰箱	réfrigérateur à groupe incorporé
5.1-39	soda-fountain ◐	冷饮柜台	fontaine à soda
5.1-40	storage shelf area	搁板存放面积	surface utile de rangement
5.1-41	three-star section	三星级冻结室	partie "trois étoiles"
5.1-42	top-opening type	顶开式	appareil du type coffre
5.1-43	two-star section	两星区	partie "deux étoiles"
5.1-44	upright freezer	立式家用冷冻柜	congélateur vertical *congélateur-armoire*
5.1-45	upright type	立式	appareil du type armoire
5.1-46	vitreous enamel	搪瓷	émail vitrifié
5.1-47	water dispenser ○ *drinking water cooler* *fountain (USA)* ◐	饮用水冷却器	rafraîchisseur d'eau (potable) *fontaine réfrigérée*

	SECTION 5.2 *Commercial equipment*	节 **5.2** 商用设备	**SOUS-CHAPITRE 5.2** *Equipements commerciaux*
5.2-**1**	air curtain	空气幕	rideau d'air
5.2-**2**	air outlet	出风口	soufflage d'air
5.2-**3**	air return	回风口	reprise d'air
5.2-**4**	assisted-service cabinet	辅助陈列柜	meuble à service assisté
5.2-**5**	automatic food-vending machine	自动食品售货机	distributeur automatique de denrées
5.2-**6**	base	基面	plan d'exposition inférieur
5.2-**7**	cabinet shell	冷柜外壳	enveloppe (d'un meuble frigorifique) *cuve extérieure (d'un meuble frigorifique)* ◐

ENGLISH	汉语	FRANÇAIS	
canopy	盖板	fronton	5.2-**8**
closed refrigerated display cabinet	闭式冷藏陈列柜	meuble frigorifique de présentation fermé	5.2-**9**
commercial cabinet *commercial refrigerator* ◐ *reach-in refrigerator* ◐ *service cabinet* ◐	商用冷柜	armoire frigorifique commerciale *réfrigérateur commercial*	5.2-**10**
commercial refrigerator ◐ *commercial cabinet* *reach-in refrigerator* ◐ *service cabinet* ◐	商用冷柜	armoire frigorifique commerciale *réfrigérateur commercial*	5.2-**11**
counter ◐ *display cabinet*	陈列柜	meuble de vente *comptoir* *meuble d'étalage* *meuble d'exposition* *présentoir* *vitrine* ◐	5.2-**12**
display *display opening*	陈列 展示陈列	exposition *étalage* *zone d'exposition* *zone de présentation*	5.2-**13**
display cabinet *counter* ◐	陈列柜	meuble de vente *comptoir* *meuble d'étalage* *meuble d'exposition* *présentoir* *vitrine* ◐	5.2-**14**
display opening *display*	陈列 展示陈列	exposition *étalage* *zone d'exposition* *zone de présentation*	5.2-**15**
display opening area	展示陈列区	surface de la zone d'exposition	5.2-**16**
end cabinet	末端陈列柜	meuble tête de gondole	5.2-**17**
end wall	端板	joue *panneau d'extrémité*	5.2-**18**
external angle	外角	meuble d'angle ouvert	5.2-**19**
front	前面板	face avant	5.2-**20**
front riser	前部提升机	arrêt produit *arrêtoir*	5.2-**21**
frozen-food cabinet	冻结食品柜	conservateur pour produits congelés	5.2-**22**
frozen-food display case ◐	冻结食品陈列柜	comptoir pour produits congelés	5.2-**23**
handrail	扶手	main courante	5.2-**24**
horizontal refrigerated display cabinet	卧式冷藏陈列柜	meuble frigorifique de vente horizontal (MFV horizontal)	5.2-**25**
internal angle	内角	meuble d'angle fermé	5.2-**26**
internal gross volume	内部毛容积	volume intérieur brut	5.2-**27**
internal net volume	内净容积	volume intérieur net	5.2-**28**
island cabinet *island display case* ◐	岛式柜 岛式陈列柜	gondole *meuble îlot*	5.2-**29**
island display case ◐ *island cabinet*	岛式柜 岛式陈列柜	gondole *meuble îlot*	5.2-**30**
island run	岛式布置	îlot	5.2-**31**

	ENGLISH	汉语	FRANÇAIS
5.2-**32**	kickplate	踢脚板	plinthe
5.2-**33**	line-up	排列整齐	linéaire
5.2-**34**	load limit	负载极限	limite de chargement
5.2-**35**	load limit line	负载界线	ligne de limite de chargement
5.2-**36**	night cover	夜间罩	protecteur de nuit *couvercle de nuit*
5.2-**37**	normal conditions of use	标准使用工况	conditions normales d'emploi
5.2-**38**	open refrigerated display cabinet	开式冷藏陈列柜	meuble frigorifique ouvert
5.2-**39**	overall external dimensions at installation	安装总外形尺寸	encombrement hors-tout à l'installation
5.2-**40**	overall external dimensions in service	使用总外形尺寸	encombrement hors-tout en service
5.2-**41**	price-marking rail	价牌挂杆	porte-étiquettes
5.2-**42**	reach-in refrigerator ◐ *commercial cabinet* *commercial refrigerator* ◐ *service cabinet* ◐	商用冷柜	armoire frigorifique commerciale *réfrigérateur commercial*
5.2-**43**	refrigerated cabinet ○	冷藏柜	armoire frigorifique *meuble frigorifique*
5.2-**44**	refrigerated counter ◐ *refrigerated display cabinet* *refrigerated display case* ◐ *refrigerated showcase* ◐	冷藏陈列柜	meuble frigorifique de vente (MFV) *vitrine frigorifique* *présentoir frigorifique* ◐ *comptoir frigorifique* ○
5.2-**45**	refrigerated display cabinet *refrigerated counter* ◐ *refrigerated display case* ◐ *refrigerated showcase* ◐	冷藏陈列柜	meuble frigorifique de vente (MFV) *vitrine frigorifique* *présentoir frigorifique* ◐ *comptoir frigorifique* ○
5.2-**46**	refrigerated display case ◐ *refrigerated display cabinet* *refrigerated counter* ◐ *refrigerated showcase* ◐	冷藏陈列柜	meuble frigorifique de vente (MFV) *vitrine frigorifique* *présentoir frigorifique* ◐ *comptoir frigorifique* ○
5.2-**47**	refrigerated shelf area	冷藏搁架面积	surface des étagères d'un meuble frigorifique
5.2-**48**	refrigerated showcase ◐ *refrigerated display cabinet* *refrigerated counter* ◐ *refrigerated display case* ◐	冷藏陈列柜	meuble frigorifique de vente (MFV) *vitrine frigorifique* *présentoir frigorifique* ◐ *comptoir frigorifique* ○
5.2-**49**	refrigerated window	冷藏橱窗	devanture frigorifique *vitrine frigorifique*
5.2-**50**	retro-case	立式冷藏柜	meuble à service arrière
5.2-**51**	roll-in cabinet	滚入式冷藏柜	meuble à façade levable
5.2-**52**	roll-in refrigerator	滚入式冷藏间	armoire frigorifique à chariots
5.2-**53**	self-service cabinet *self-service display case* ◐	自助式冷藏柜 自助式陈列柜	meuble libre-service *comptoir libre-service*
5.2-**54**	self-service display case ◐ *self-service cabinet*	自助式冷藏柜 自助式陈列柜	meuble libre-service *comptoir libre-service*
5.2-**55**	semi-vertical refrigerated display cabinet	半立式冷藏展示柜	meuble frigorifique de vente semi-vertical
5.2-**56**	service cabinet ◐ *commercial cabinet* *commercial refrigerator* ◐ *reach-in refrigerator* ◐	商用冷柜	armoire frigorifique commerciale *réfrigérateur commercial*

ENGLISH	汉语	FRANÇAIS	
shelf	搁板	étagère	5.2-**57**
shelf sham	假搁板	fausse masse	5.2-**58**
storage volume	贮藏容积	volume utile	5.2-**59**
technical line-up	技术排列	linéaire technique	5.2-**60**
vertical refrigerated display cabinet	立式冷藏陈列柜	meuble frigorifique de vente vertical	5.2-**61**

SECTION 5.3 *Cold rooms*	节 **5.3** 冷藏间	**SOUS-CHAPITRE 5.3** *Chambres froides*	
air lock	门斗	sas d'air	5.3-**1**
anteroom	川堂	antichambre	5.3-**2**
capacity of a cold room	冷房间容量	capacité d'une chambre froide	5.3-**3**
chill room ◐ *chilling room* *cooler (USA)* ◐	冷却间	chambre de réfrigération	5.3-**4**
chilling room *chill room* ◐ *cooler (USA)* ◐	冷却间	chambre de réfrigération	5.3-**5**
cold chamber ◐ *cold room*	冷房间	chambre froide	5.3-**6**
cold room *cold chamber* ◐	冷房间	chambre froide	5.3-**7**
cold storage room	冷藏间	chambre d'entreposage frigorifique *chambre de stockage frigorifique* ◐	5.3-**8**
cooler (USA) ◐ *chilling room* *chill room* ◐	冷却间	chambre de réfrigération	5.3-**9**
cooling tunnel	冷却隧道	tunnel de réfrigération	5.3-**10**
effective capacity ◐ *net capacity*	净容量	volume utile d'une chambre froide	5.3-**11**
freezer room ◐ *frozen-food storage room*	冻结物冷藏间	chambre (d'entreposage) pour produits congelés	5.3-**12**
frozen-food storage room *freezer room* ◐	冻结物冷藏间	chambre (d'entreposage) pour produits congelés	5.3-**13**
gross capacity *gross volume* ◐	毛容量	volume brut d'une chambre froide	5.3-**14**
gross volume ◐ *gross capacity*	毛容量	volume brut d'une chambre froide	5.3-**15**
inflatable cold store	充气膨胀式冷藏库	chambre froide gonflable	5.3-**16**
jacketed cold room	夹套式冷间	chambre (froide) à double paroi *chambre (froide) à enveloppe d'air froid*	5.3-**17**
net capacity *effective capacity* ◐	净容量	volume utile d'une chambre froide	5.3-**18**
portable cold room	移动式冷房间	chambre froide transportable	5.3-**19**
precooling room	预冷间	chambre de préréfrigération	5.3-**20**
sectional cold room	装配式冷房间	chambre froide préfabriquée	5.3-**21**

	ENGLISH	汉语	FRANÇAIS
5.3-**22**	store contents	贮藏容量	chargement d'une chambre
5.3-**23**	vestibule with doors	带门的门厅	antichambre
5.3-**24**	walk-in freezer ○	小型冻结物冷藏间	congélateur-chambre ○
5.3-**25**	walk-in refrigerator ○	小型冷藏间	réfrigérateur-chambre ○

	SECTION 5.4 *Cold stores*	节 **5.4** 冷藏库	**SOUS-CHAPITRE 5.4** *Entrepôts frigorifiques*
5.4-**1**	bulk storage *storage in bulk* *bulk stowage* ◐	散装贮藏	entreposage en vrac
5.4-**2**	bulk stowage ◐ *bulk storage* *storage in bulk*	散装贮藏	entreposage en vrac
5.4-**3**	cold storage *refrigerated storage*	冷藏	entreposage frigorifique
5.4-**4**	cold store *refrigerated warehouse*	冷藏库	entrepôt frigorifique
5.4-**5**	cold store combine ◐ *cold store facility* *cold store complex* ◐	冷藏联合企业	complexe frigorifique
5.4-**6**	cold store complex ◐ *cold store facility* *cold store combine* ◐	冷藏联合企业	complexe frigorifique
5.4-**7**	cold store facility *cold store combine* ◐ *cold store complex* ◐	冷藏联合企业	complexe frigorifique
5.4-**8**	despatching cold store	分配性冷藏库	entrepôt frigorifique de distribution
5.4-**9**	dock leveller	月台跳板	dispositif de mise à niveau (de quai)
5.4-**10**	engine house	机房	bâtiment des machines
5.4-**11**	frozen-food store	冻结物冷藏库	entrepôt pour produits congelés
5.4-**12**	gangway ○ *passage*	过道	allée *passage* ◐
5.4-**13**	high-rise store	高架冷藏库	entrepôt de grande hauteur
5.4-**14**	loading bank ○ *platform*	装卸月台	quai
5.4-**15**	maximum stacking density	最大堆放密度	densité de gerbage
5.4-**16**	multistorey cold store ◐	多层冷藏库	entrepôt frigorifique à étages
5.4-**17**	multipurpose cold store	多用途冷藏库	entrepôt frigorifique polyvalent
5.4-**18**	nursery cold store	苗木冷藏库	chambre froide à plants
5.4-**19**	order-picking area	配货区	zone de préparation des commandes
5.4-**20**	palletizing area	托盘堆放区	aire de palettisation
5.4-**21**	passage *gangway* ○	过道	allée *passage* ◐
5.4-**22**	platform *loading bank* ○	装卸月台	quai

ENGLISH	汉语	FRANÇAIS	
platform height floor *truck height floor* ◐	高地坪	plancher bas à niveau de quai	5.4-**23**
port cold store	港口冷藏库	entrepôt frigorifique portuaire	5.4-**24**
port door	装卸门	porte d'accostage	5.4-**25**
private cold store	私用冷藏库	entrepôt frigorifique privé	5.4-**26**
public cold store	公用冷藏库	entrepôt frigorifique public	5.4-**27**
ramp *steep* ◐	坡道	rampe d'accès	5.4-**28**
refrigerated storage *cold storage*	冷藏	entreposage frigorifique	5.4-**29**
refrigerated warehouse *cold store*	冷藏库	entrepôt frigorifique	5.4-**30**
shelf storage *shelf stowage* ◐	搁架式贮藏	entreposage sur étagères	5.4-**31**
shelf stowage ◐ *shelf storage*	搁架式贮藏	entreposage sur étagères	5.4-**32**
single-storey cold store ◐	单层冷藏库	entrepôt frigorifique à un seul niveau	5.4-**33**
slab-on-ground floor	低地坪	plancher bas à niveau de sol	5.4-**34**
specialized cold store	专用冷藏库	entrepôt frigorifique spécialisé	5.4-**35**
stacking allotment	成批堆垛	allotissement *lotissement*	5.4-**36**
stacking density	堆垛密度	densité maximale d'entreposage	5.4-**37**
steep ◐ *ramp*	坡道	rampe d'accès	5.4-**38**
stock turnover	贮藏周转库	rotation des stocks	5.4-**39**
storage factor	贮藏系数	taux de remplissage (d'un entrepôt)	5.4-**40**
storage in bulk *bulk storage* *bulk stowage* ◐	散装贮藏	entreposage en vrac	5.4-**41**
store *warehouse*	贮藏	entrepôt	5.4-**42**
truck height floor ◐ *platform height floor*	高地坪	plancher bas à niveau de quai	5.4-**43**
warehouse *store*	贮藏	entrepôt	5.4-**44**
warming room	加热（升温）间 升温间	salle de réchauffage	5.4-**45**

SECTION 5.5 *Refrigerated transport* **SECTION 5.5.1** *Refrigerated transport special technology*	节 **5.5** 冷藏运输 节 **5.5.1** 冷藏运输专门技术	**SOUS-CHAPITRE 5.5** *Transport frigorifique* **SOUS-CHAPITRE 5.5.1** *Techniques spécifiques au transport frigorifique*	
aeolian fan *wind-turbine fan*	风力风机	ventilateur à turbine éolienne	5.5.1-**1**
air leakage	漏气	fuite d'air	5.5.1-**2**

	ENGLISH	汉语	FRANÇAIS
5.5.1-**3**	ATP agreement	ATP 协议	accord ATP
5.5.1-**4**	body-icing	夹冰货物	glaçage direct au sein du chargement
5.5.1-**5**	combined transport	联合运输	transport combiné
5.5.1-**6**	contact icing	接触加冰	glaçage direct (des denrées)
5.5.1-**7**	controlled-atmosphere transport	气调运输	transport sous atmosphère contrôlée
5.5.1-**8**	controlled-temperature transport	控温运输	transport sous température dirigée
5.5.1-**9**	dry-ice bunker	干冰车厢	compartiment à glace carbonique
5.5.1-**10**	expendable refrigerant system	不回收式制冷系统	équipement à frigorigène perdu
5.5.1-**11**	ice blower *ice gun*	风力送冰机	lance-glace
5.5.1-**12**	ice bunker	冰仓	bac à glace *panier à glace*
5.5.1-**13**	ice-bunker screen	冰仓隔板	écran de panier à glace
5.5.1-**14**	ice gun *ice blower*	风力送冰机	lance-glace
5.5.1-**15**	icing hatch	加冰口	trappe de chargement de la glace
5.5.1-**16**	insulated body	保温车体	caisse isolée *caisse isotherme* ○
5.5.1-**17**	insulated vehicle	保温车	véhicule isolé *véhicule isotherme* ○
5.5.1-**18**	interface	交接面	interface
5.5.1-**19**	intermodal transport	联合运输	transport intermodal
5.5.1-**20**	land transport equipment	陆上运输工具	engin de transport terrestre
5.5.1-**21**	mechanically refrigerated vehicle	机械冷藏车	véhicule frigorifique
5.5.1-**22**	multimodal transport	多式联运	transport multimodal
5.5.1-**23**	over-the-road system ◐	车装制冷系统	système de refroidissement en route ◐
5.5.1-**24**	piggy-back transport	车背式运输	ferroutage *transport rail-route* ◐
5.5.1-**25**	re-icing	补充冰	reglaçage
5.5.1-**26**	refrigerated transport	冷藏运输	transport frigorifique
5.5.1-**27**	refrigerated vehicle	冷藏车	véhicule réfrigérant
5.5.1-**28**	top-icing	顶部加冰	glaçage direct sur le chargement
5.5.1-**29**	ventilated vehicle	通风车	véhicule aéré
5.5.1-**30**	wind-turbine fan *aeolian fan*	风力风机	ventilateur à turbine éolienne

	SECTION 5.5.2 *Road transport*	节 **5.5.2** 公路运输	**SOUS-CHAPITRE 5.5.2** *Transport routier*
5.5.2-**1**	air-flow floor *rack floor*	通风底板	plancher soufflant *plancher ventilé*
5.5.2-**2**	curtain-sider	侧帘	remorque à bâches latérales amovibles
5.5.2-**3**	delivery vehicle	配送车辆	véhicule de livraison isotherme
5.5.2-**4**	distribution hub	配送中心	plate-forme de distribution

ENGLISH	汉语	FRANÇAIS	
fuel-freeze system	液化石油气系统	équipement à frigorigène courant	5.5.2-**5**
insulated large van *insulated lorry* *insulated truck (USA)*	保温汽车	camion isolé	5.5.2-**6**
insulated lorry *insulated truck (USA)* *insulated large van*	保温汽车	camion isolé	5.5.2-**7**
insulated truck (USA) *insulated lorry* *insulated large van*	保温汽车	camion isolé	5.5.2-**8**
mechanically refrigerated large van *mechanically refrigerated lorry* *mechanically refrigerated truck (USA)*	机械冷藏车 机械冷藏汽车	camion frigorifique	5.5.2-**9**
mechanically refrigerated lorry *mechanically refrigerated large van* *mechanically refrigerated truck (USA)*	机械冷藏车 机械冷藏汽车	camion frigorifique	5.5.2-**10**
mechanically refrigerated truck (USA) *mechanically refrigerated lorry* *mechanically refrigerated large van*	机械冷藏车 机械冷藏汽车	camion frigorifique	5.5.2-**11**
multicompartment vehicle	多仓运输汽车	véhicule multicompartiment	5.5.2-**12**
nose-mounted refrigeration unit	车首制冷机组	groupe frigorifique frontal	5.5.2-**13**
rack floor *air-flow floor*	通风底板	plancher soufflant *plancher ventilé*	5.5.2-**14**
refrigerated large van *refrigerated lorry* *refrigerated truck (USA)*	冷藏汽车	camion réfrigérant	5.5.2-**15**
refrigerated lorry *refrigerated large van* *refrigerated truck (USA)*	冷藏汽车	camion réfrigérant	5.5.2-**16**
refrigerated truck (USA) *refrigerated lorry* *refrigerated large van*	冷藏汽车	camion réfrigérant	5.5.2-**17**
road tanker ◐ *tank truck (USA)* ◐	公路槽车	camion citerne	5.5.2-**18**
tank truck (USA) ◐ *road tanker* ◐	公路槽车	camion citerne	5.5.2-**19**
trailer	拖车	remorque	5.5.2-**20**
underslung refrigeration unit	车底制冷机组	groupe frigorifique sous châssis	5.5.2-**21**

SECTION 5.5.3 *Rail transport*	节 **5.5.3** 铁路运输	**SOUS-CHAPITRE 5.5.3** *Transport ferroviaire*	
block train	编组冷藏列车	train bloc	5.5.3-**1**
end-bunker refrigerated railcar (USA) ◐ *end-bunker refrigerated truck* ◐	端式冰仓铁路冷藏车	wagon réfrigérant à bac d'extrémité ◐	5.5.3-**2**
end-bunker refrigerated truck ◐ *end-bunker refrigerated railcar (USA)* ◐	端式冰仓铁路冷藏车	wagon réfrigérant à bac d'extrémité ◐	5.5.3-**3**
iced railcar (USA) *iced truck*	铁路加冰车	wagon glacé	5.5.3-**4**
iced truck *iced railcar (USA)*	铁路加冰车	wagon glacé	5.5.3-**5**

	ENGLISH	汉语	FRANÇAIS
5.5.3-**6**	icing of railcars (USA) *icing of trucks*	铁路车辆加冰	glaçage des wagons
5.5.3-**7**	icing of trucks *icing of railcars (USA)*	铁路车辆加冰	glaçage des wagons
5.5.3-**8**	icing platform	加冰台	plate-forme de glaçage
5.5.3-**9**	icing tower	加冰塔	tour de glaçage
5.5.3-**10**	insulated railcar (USA) *insulated railway truck* *insulated truck*	铁路保温车	wagon isolé
5.5.3-**11**	insulated railway truck *insulated truck* *insulated railcar (USA)*	铁路保温车	wagon isolé
5.5.3-**12**	insulated truck *insulated railway truck* *insulated railcar (USA)*	铁路保温车	wagon isolé
5.5.3-**13**	mechanically refrigerated railcar (USA) *mechanically refrigerated truck* *refrigerator car (USA)*	铁路机械冷藏车	wagon frigorifique
5.5.3-**14**	mechanically refrigerated truck *mechanically refrigerated railcar (USA)* *refrigerator car (USA)*	铁路机械冷藏车	wagon frigorifique
5.5.3-**15**	overhead-bunker refrigerated railcar (USA) *overhead-bunker refrigerated truck*	顶式冰仓铁路冷藏车	wagon réfrigérant à bac plafonnier
5.5.3-**16**	overhead-bunker refrigerated truck *overhead-bunker refrigerated railcar (USA)*	顶式冰仓铁路冷藏车	wagon réfrigérant à bac plafonnier
5.5.3-**17**	rail tanker ◐	铁路槽车	wagon citerne ◐
5.5.3-**18**	refrigerated railcar (USA) *refrigerated truck*	铁路冷藏车	wagon réfrigérant
5.5.3-**19**	refrigerated train	冷藏列车	train frigorifique
5.5.3-**20**	refrigerated truck *refrigerated railcar (USA)*	铁路冷藏车	wagon réfrigérant
5.5.3-**21**	refrigerator car (USA) *mechanically refrigerated truck* *mechanically refrigerated railcar (USA)*	铁路机械冷藏车	wagon frigorifique

	SECTION 5.5.4 *Marine transport*	节 **5.5.4** 船舶运输	**SOUS-CHAPITRE 5.5.4** *Transport maritime*
5.5.4-**1**	barge carrier	载驳船	navire porte-barges *navire porte-chalands*
5.5.4-**2**	bottom plating	船底衬板	vaigrage de fond
5.5.4-**3**	bulkhead	舱壁	cloison
5.5.4-**4**	cargo door *hull loading door*	货门 (舷侧)货门	portelone *sabord de charge*
5.5.4-**5**	cells (of a container ship)	(集装箱船)箱位	cellules (de navires porte-conteneurs)
5.5.4-**6**	cellular (container) ship	井架式(集装箱)船	navire (porte-conteneurs) cellulaire
5.5.4-**7**	(container) guides	(集装箱)导架	guides (de conteneurs)
5.5.4-**8**	container ship *container vessel*	集装箱船	navire porte-conteneurs

ENGLISH	汉语	FRANÇAIS	
container vessel *container ship*	集装箱船	navire porte-conteneurs	5.5.4-**9**
deck	甲板	pont	5.5.4-**10**
dry cargo	干货	marchandises ordinaires	5.5.4-**11**
factory ship	加工船	navire-usine	5.5.4-**12**
freezing trawler	冷冻拖网渔船	chalutier congélateur	5.5.4-**13**
gas-carrier	气体运输船	navire gazier	5.5.4-**14**
hatch ◐	舱口盖	écoutille *panneau (de cale)* ◐	5.5.4-**15**
hold	货舱	cale	5.5.4-**16**
hold pen ◐ *hold pound*	鱼舱	compartiment de cale	5.5.4-**17**
hold pound *hold pen* ◐	鱼舱	compartiment de cale	5.5.4-**18**
hull	船壳	coque	5.5.4-**19**
hull loading door *cargo door*	货门 (舷侧)货门	portelone *sabord de charge*	5.5.4-**20**
hull plating *planking* *shell plating*	船壳板	bordé extérieur	5.5.4-**21**
inner lining *inner planking* ◐ *inner plating* ◐	内衬板	vaigrage	5.5.4-**22**
inner planking ◐ *inner lining* *inner plating* ◐	内衬板	vaigrage	5.5.4-**23**
inner plating ◐ *inner lining* *inner planking* ◐	内衬板	vaigrage	5.5.4-**24**
integrated tank ◐ *membrane tank* ◐	薄膜罐	cuve à membrane *cuve intégrée*	5.5.4-**25**
marine refrigerating plant	船用制冷装置	installation frigorifique marine	5.5.4-**26**
membrane tank ◐ *integrated tank* ◐	薄膜罐	cuve à membrane *cuve intégrée*	5.5.4-**27**
methane carrier *methane tanker*	甲烷运输船	(navire) méthanier	5.5.4-**28**
methane tanker *methane carrier*	甲烷运输船	(navire) méthanier	5.5.4-**29**
mother ship	母船	navire-base	5.5.4-**30**
planking *hull plating* *shell plating*	船壳板	bordé extérieur	5.5.4-**31**
reefer ship *refrigerated cargo vessel*	冷藏船	navire frigorifique	5.5.4-**32**
refrigerated (cargo) hold	冷藏货舱	cale refroidie	5.5.4-**33**
refrigerated cargo vessel *reefer ship*	冷藏船	navire frigorifique	5.5.4-**34**
roll-on roll-off ship	滚装船	roulier *transroulier* ◐	5.5.4-**35**

	ENGLISH	汉语	FRANÇAIS
5.5.4-**36**	self-supporting tank ◐ *structural tank*	自承罐	cuve autoporteuse
5.5.4-**37**	shell plating *hull plating* *planking*	船壳板	bordé extérieur
5.5.4-**38**	stanchion	支柱	épontille ◐
5.5.4-**39**	structural tank *self-supporting tank* ◐	自承罐	cuve autoporteuse
5.5.4-**40**	tweendeck	甲板间	entrepont

	SECTION 5.5.5 *Air transport*	节 **5.5.5** 空运	**SOUS-CHAPITRE 5.5.5** *Transport aérien*
5.5.5-**1**	air cargo	空运货物	fret aérien
5.5.5-**2**	air carrier	空运机构	transporteur aérien
5.5.5-**3**	air container *ULD (Unit Load Device)*	单位装载设备	ULD (unité de charge) *conteneur aérien*
5.5.5-**4**	air freight	空运货物	fret aérien
5.5.5-**5**	anti-icer	防冰器	anti-givre
5.5.5-**6**	anti-icing	防冰	anti-givrage
5.5.5-**7**	belly cargo *belly hold*	下部货舱	fret soute inférieure
5.5.5-**8**	belly hold *belly cargo*	下部货舱	fret soute inférieure
5.5.5-**9**	clearance	间隙	espace libre
5.5.5-**10**	combi aircraft	客货飞机	avion combi
5.5.5-**11**	ground equipment	地面设备	équipement au sol
5.5.5-**12**	hub	中枢	aéroport principal *plaque tournante*
5.5.5-**13**	igloo	园顶集装箱	conteneur igloo
5.5.5-**14**	insulating ULD *thermal ULD*	隔热集装箱	unité de charge isolée
5.5.5-**15**	landing weight	着陆重量	masse à l'atterrissage
5.5.5-**16**	loose cargo	散装货	fret en vrac
5.5.5-**17**	lower deck *lower hold*	底舱	pont inférieur
5.5.5-**18**	lower hold *lower deck*	底舱	pont inférieur
5.5.5-**19**	main deck	主舱	pont principal
5.5.5-**20**	PAG *pallet*	托盘	palette
5.5.5-**21**	pallet *PAG*	托盘	palette
5.5.5-**22**	payload	有效载荷	charge marchande
5.5.5-**23**	perishable cargo	易腐货物	fret périssable
5.5.5-**24**	pressurization	增压舱	pressurisation

ENGLISH	汉语	FRANÇAIS	
ramp weight	起飞重量	masse au stationnement	5.5.5-**25**
thermal ULD *insulating ULD*	隔热集装箱	unité de charge isolée	5.5.5-**26**
turnaround	转向	temps nécessaire au déchargement et au chargement	5.5.5-**27**
ULD (Unit Load Device) *air container*	单位装载设备	ULD (unité de charge) *conteneur aérien*	5.5.5-**28**
widebody aircraft	宽体飞机	avion gros porteur	5.5.5-**29**

SECTION 5.6 *Handling equipment*	节 **5.6** 处理设备	**SOUS-CHAPITRE 5.6** *Matériel de manutention*	
box pallet *bulk bin* *pallet bin* *pallet box* *pallet crate* *palletainer* *pallox*	箱式托盘	caisse palette *pallox* ◐	5.6-**1**
bulk bin *box pallet* *pallet bin* *pallet box* *pallet crate* *palletainer* *pallox*	箱式托盘	caisse palette *pallox* ◐	5.6-**2**
bulkhead	隔仓	cloison "écran"	5.6-**3**
bulking *grouping*	散装	groupage	5.6-**4**
ceiling air duct	顶棚风道	conduit de ventilation au plafond	5.6-**5**
clip-on unit	快接式制冷机组	groupe frigorifique amovible	5.6-**6**
converter bar *pallet converter* *pallet posts*	托盘支柱	convertisseur (pour palettes)	5.6-**7**
depalletize (to)	卸货（动词）	dépalettiser	5.6-**8**
dolly	滚轮小车	support à roulettes	5.6-**9**
europallet	欧式托盘	europalette	5.6-**10**
floor air duct	地板风道	conduit sous plancher	5.6-**11**
forklift truck	叉式码垛车	chariot (élévateur) à fourche	5.6-**12**
(freight) container *(transport) container*	集装箱 货运集装箱	conteneur (de transport) *container* ○	5.6-**13**
grouping *bulking*	散装	groupage	5.6-**14**
handling	搬运	manutention	5.6-**15**
heated container	加热集装箱	conteneur chauffé	5.6-**16**
insulated container	隔热集装箱	conteneur isolé	5.6-**17**
ISO container	国际标准集装箱	conteneur ISO	5.6-**18**
lift truck	码垛车	chariot élévateur	5.6-**19**

	ENGLISH	汉语	FRANÇAIS
5.6-**20**	mechanically refrigerated container	机械制冷集装箱	conteneur frigorifique
5.6-**21**	pallet	托盘	palette
5.6-**22**	pallet bin *box pallet* *bulk bin* *pallet box* *pallet crate* *palletainer* *pallox*	箱式托盘	caisse palette *pallox* ◐
5.6-**23**	pallet box *box pallet* *bulk bin* *pallet bin* *pallet crate* *palletainer* *pallox*	箱式托盘	caisse palette *pallox* ◐
5.6-**24**	pallet converter *converter bar* *pallet posts*	托盘支柱	convertisseur (pour palettes)
5.6-**25**	pallet crate *box pallet* *bulk bin* *pallet bin* *pallet box* *palletainer* *pallox*	箱式托盘	caisse palette *pallox* ◐
5.6-**26**	pallet loader *pallet truck* *pedestrian forklift*	托盘搬运车	transpalette
5.6-**27**	pallet posts *converter bar* *pallet converter*	托盘支柱	convertisseur (pour palettes)
5.6-**28**	pallet rack	托盘架	palettier
5.6-**29**	pallet truck *pallet loader* *pedestrian forklift*	托盘搬运车	transpalette
5.6-**30**	palletainer *box pallet* *bulk bin* *pallet bin* *pallet box* *pallet crate* *pallox*	箱式托盘	caisse palette *pallox* ◐
5.6-**31**	palletization	托盘化	palettisation
5.6-**32**	palletize (to)	装托盘（动词） 装托盘	palettiser
5.6-**33**	pallox *box pallet* *bulk bin* *pallet bin* *pallet box* *pallet crate* *palletainer*	箱式托盘	caisse palette *pallox* ◐
5.6-**34**	pedestrian forklift *pallet loader* *pallet truck*	托盘搬运车	transpalette
5.6-**35**	pile (to) *stack (to)*	堆垛（动词）	gerber *empiler*

ENGLISH	汉语	FRANÇAIS	
porthole container	外接式保温集装箱	conteneur à hublots	5.6-**36**
reach truck	伸臂叉车	chariot (élévateur) à fourche rétractable *chariot élévateur à mât rétractable* ◐	5.6-**37**
reefer container *refrigerated container*	冷藏集装箱	conteneur réfrigéré	5.6-**38**
refrigerated container *reefer container*	冷藏集装箱	conteneur réfrigéré	5.6-**39**
refrigerated and heated container	冷热两用集装箱	conteneur réfrigéré et chauffé	5.6-**40**
refrigerated and heated container with controlled or modified atmosphere	冷热气调集装箱	conteneur réfrigéré et chauffé à atmosphère contrôlée	5.6-**41**
removable equipment	可拆卸设备	équipement amovible *unité amovible* ◐	5.6-**42**
roll container	滚装集装箱	roll conteneur	5.6-**43**
stack (to) *pile (to)*	堆垛（动词）	gerber *empiler*	5.6-**44**
stack height	堆垛高度	hauteur de gerbage	5.6-**45**
stacking	堆装	gerbage	5.6-**46**
swap body	交换箱体	conteneur rail-route	5.6-**47**
tank container	罐式集装箱	conteneur-citerne	5.6-**48**
test pressure	试验压力	pression d'épreuve	5.6-**49**
thermal container	高温集装箱	conteneur à caractéristiques thermiques	5.6-**50**
(transport) container *(freight) container*	集装箱 货运集装箱	conteneur (de transport) *container* ○	5.6-**51**

章 6. 易腐品的冷藏

◐ 容许（或默许）术语

○ 过时术语

ENGLISH	汉语	FRANÇAIS	
SECTION 6.1 *Food characteristics*	节 **6.1** 食品特性	**SOUS-CHAPITRE 6.1** *Caractéristiques des aliments*	
aroma	芳香	arôme *parfum* ◐	6.1-**1**
boil-in-bag food	袋煮食品	aliment précuit en sachet	6.1-**2**
bound water	结合水	eau liée	6.1-**3**
chilled food	冷却食品	denrées réfrigérées	6.1-**4**
cloudiness	浑浊	turbidité *louche* ◐ *trouble* ◐	6.1-**5**
convenience food	方便食品	aliment prêt à l'emploi	6.1-**6**
fermentation heat	发酵热	chaleur de fermentation	6.1-**7**
flavour	风味	flaveur	6.1-**8**
food	食品	aliment	6.1-**9**
free water	游离水	eau libre	6.1-**10**
freeze-thaw resistance	耐冻-融性	résistance à la congélation-décongélation	6.1-**11**
institutional food	团体份饭	aliment pour collectivité	6.1-**12**
keeping quality	耐藏性	aptitude à la conservation	6.1-**13**
keeping time ○ *storage time*	贮存时间	durée de conservation *durée d'entreposage* ◐	6.1-**14**
metabolic heat	代谢热	chaleur métabolique	6.1-**15**
minimally prepared food	保鲜简装	produit alimentaire de 4e gamme	6.1-**16**
organoleptic qualities	感观品质	qualités organoleptiques *qualités sensorielles* ◐	6.1-**17**
practical storage life	实际贮存寿命	durée pratique de conservation *durée pratique d'entreposage*	6.1-**18**
precooked dish ◐ *prepared food*	预制食品	aliment cuisiné *aliment précuit* *plat préparé* ◐	6.1-**19**
prepared food *precooked dish* ◐	预制食品	aliment cuisiné *aliment précuit* *plat préparé* ◐	6.1-**20**
quality assessment *quality evaluation* ◐	质量评定	évaluation de la qualité *appréciation de la qualité* ◐	6.1-**21**
quality control	质量控制	contrôle de la qualité *maintien de la qualité* *maîtrise de la qualité*	6.1-**22**
quality evaluation ◐ *quality assessment*	质量评定	évaluation de la qualité *appréciation de la qualité* ◐	6.1-**23**
rate of respiration *respiratory intensity* ◐	呼吸率 呼吸强度	intensité respiratoire	6.1-**24**
respiratory heat	呼吸热	chaleur de respiration	6.1-**25**
respiratory intensity ◐ *rate of respiration*	呼吸率 呼吸强度	intensité respiratoire	6.1-**26**
sensory (organoleptic) properties	（传入感觉器官）感觉性质	propriétés organoleptiques *propriétés sensorielles* ◐	6.1-**27**

	ENGLISH	汉语	FRANÇAIS
6.1-28	storage time *keeping time* ○	贮存时间	durée de conservation *durée d'entreposage* ◐
6.1-29	sugar content	含糖量	teneur en sucre
6.1-30	taste	滋味	goût
6.1-31	time-temperature tolerance (TTT)	时间-温度允许度	corrélation temps-température *tolérance temps-température* ◐
6.1-32	tolerance	允许度	tolérance

	SECTION 6.2 *Food deterioration and hygiene*	节 **6.2** 食品变质与卫生	**SOUS-CHAPITRE 6.2** *Altération des aliments et hygiène alimentaire*
6.2-1	acceptance criteria	验收准则	critère d'acceptabilité
6.2-2	acceptance level	验收等级	niveau d'acceptabilité
6.2-3	adverse influence	有害影响	altération *détérioration* ◐
6.2-4	bacterial decay ◐ *bacterial deterioration*	细菌性变质	altération bactérienne
6.2-5	bacterial deterioration *bacterial decay* ◐	细菌性变质	altération bactérienne
6.2-6	chilling injury	冷却损伤	accident causé par la réfrigération
6.2-7	clean	清洁	propre
6.2-8	cleanable	可清洁的	nettoyable
6.2-9	cleaning	清洁（处理）	nettoyage
6.2-10	cold injury *low-temperature injury* ◐	低温损伤	altération due aux basses températures
6.2-11	contaminant	污染物	contaminant
6.2-12	contamination	污染	contamination
6.2-13	control	控制	maîtrise
6.2-14	control measures	控制措施	mesures de maîtrise
6.2-15	control (to)	控制（动词）	maîtriser
6.2-16	control point	控制点	point de maîtrise
6.2-17	corrective action	改正措施	action corrective
6.2-18	corrosion-resistant material	防腐材料	matériau résistant à la corrosion
6.2-19	crevice	裂隙	crevasse *anfractuosité* ◐
6.2-20	Critical Control Point (CCP)	临界控制点	point critique pour la maîtrise (CCP)
6.2-21	critical limit	临界条件	limite critique
6.2-22	cryophilic ○	耐冷的	cryophile
6.2-23	dead space	死区	zone non accessible
6.2-24	deterioration *spoilage*	变质	altération *détérioration* ◐
6.2-25	discolouration	变色	altération de la couleur *décoloration*
6.2-26	disinfection	消毒	désinfection

ENGLISH	汉语	FRANÇAIS	
dose-response assessment	药量-（有害）反应评估	appréciation de la relation dose-réponse	6.2-**27**
durable	耐久的	résistante	6.2-**28**
enzymatic reactions	酶反应	réactions enzymatiques	6.2-**29**
exposure assessment	曝露评估	évaluation de l'exposition (à des agents indésirables)	6.2-**30**
food area	食品表面积	zone alimentaire	6.2-**31**
food hygiene	食品卫生	hygiène des aliments *hygiène alimentaire*	6.2-**32**
food safety	食品安全性	sécurité sanitaire des aliments	6.2-**33**
food suitability	食品适用性	acceptabilité des aliments	6.2-**34**
freezer burn	冻伤	brûlure de congélation	6.2-**35**
freezing damage	冻结损坏	altération due à la congélation	6.2-**36**
freezing injury	冻伤	altération au cours de la congélation ou de l'entreposage (du produit congelé)	6.2-**37**
frost damage	冻伤	gelure	6.2-**38**
gas injury	气体损伤	intoxication gazeuse	6.2-**39**
HACCP (Hazard Analysis Critical Control Point)	HACCP（危害分析关键控制点）	HACCP (analyse des dangers et leur maîtrise aux points critiques)	6.2-**40**
HACCP plan	HACCP 计划	plan HACCP	6.2-**41**
hazard	危害性	danger	6.2-**42**
hazard analysis	危害分析	analyse des dangers	6.2-**43**
hazard assessment	危害评估	appréciation des dangers	6.2-**44**
hazard identification	危害鉴别法	identification des dangers	6.2-**45**
in-package desiccation	包内脱水	dessiccation dans les emballages	6.2-**46**
inspection	检查	contrôle	6.2-**47**
low-temperature injury ◐ *cold injury*	低温损伤	altération due aux basses températures	6.2-**48**
mechanical defect	机械损伤	altération mécanique *blessure*	6.2-**49**
moisture loss	水分损失	perte d'eau	6.2-**50**
monitoring	监控器	surveillance	6.2-**51**
non-absorbent material	无吸收能力的材料	matériau non absorbant	6.2-**52**
non-food area	非食品表面	zone non alimentaire	6.2-**53**
non-toxic material	无毒材料	matériau non toxique	6.2-**54**
oxidation	氧化	oxydation	6.2-**55**
perishable	易腐物	périssable	6.2-**56**
preventive action	预防措施	action préventive *prévention*	6.2-**57**
procedure	操作规程	procédure	6.2-**58**
psychrophilic	嗜冷的	psychrophile	6.2-**59**
psychrotrophic	向冷性	psychrotrophe	6.2-**60**
quality audit	质量检查	audit de qualité	6.2-**61**
quality control	质量控制	maîtrise de la qualité	6.2-**62**
quality insurance	质量保证	assurance qualité	6.2-**63**

	ENGLISH	汉语	FRANÇAIS
6.2-**64**	quality surveillance	质量监督	surveillance de la qualité *contrôle de la qualité*
6.2-**65**	rancidity	酉合 败	rancissement
6.2-**66**	risk	风险	risque
6.2-**67**	risk analysis	风险分析	analyse des risques
6.2-**68**	risk assessment	风险评估	appréciation des risques
6.2-**69**	risk characterization	风险特性说明	caractérisation des risques
6.2-**70**	risk communication	风险通报	communication à propos des risques
6.2-**71**	risk management	风险管理（控制）	gestion des risques
6.2-**72**	self-inspection	自检	autocontrôle
6.2-**73**	shrinkage *withering* ◐	皱缩	flétrissure
6.2-**74**	slime	粘液	couche poisseuse *couche visqueuse*
6.2-**75**	smooth	光滑	lisse *conforme*
6.2-**76**	soil	污垢	souillure
6.2-**77**	splash area	散溅面积	zone d'éclaboussures
6.2-**78**	spoilage *deterioration*	变质	altération *détérioration* ◐
6.2-**79**	storage disease	贮藏病害	maladie d'entreposage
6.2-**80**	storage disorder	储藏失调	accident d'entreposage
6.2-**81**	target level	目标水准	niveau cible
6.2-**82**	traceability	追踪能力	traçabilité
6.2-**83**	validation	确认	validation
6.2-**84**	verification	验证	vérification *contrôle*
6.2-**85**	weight loss	干耗	perte de poids *perte de masse* ◐
6.2-**86**	withering ◐ *shrinkage*	皱缩	flétrissure

	SECTION 6.3 *Refrigeration and the food industry* **SECTION 6.3.1** *Products of plant origin*	节 **6.3** 制冷与食品工业 节 **6.3.1** 植物源制品	**SOUS-CHAPITRE 6.3** *Traitement par le froid dans l'industrie agroalimentaire* **SOUS-CHAPITRE 6.3.1** *Produits végétaux*
6.3.1-**1**	abnormal external moisture	不正常的外表湿度	humidité extérieure anormale *embuage* ◐
6.3.1-**2**	accelerated ripening	促进成熟	maturation accélérée
6.3.1-**3**	acclimatized	（使）适应环境	ressuyé
6.3.1-**4**	alcohol injury	乙醇损伤	fermentation alcoolique (des fruits)
6.3.1-**5**	artificial atmosphere generator	人工气体发生器	générateur d'atmosphère

ENGLISH	汉语	FRANÇAIS	
bitter pit	苦疤	taches amères *bitter pit* ◐	6.3.1-**6**
brown core *brown heart* ◐	褐心	cœur brun	6.3.1-**7**
brown heart ◐ *brown core*	褐心	cœur brun	6.3.1-**8**
C.A. storage ◐ *controlled-atmosphere storage* *storage in controlled atmosphere* ◐	气调贮藏	entreposage en atmosphère contrôlée (AC) *entreposage AC*	6.3.1-**9**
C.A. storage room ◐ *controlled-atmosphere (storage) room* *gas store* ○	气调贮藏室	chambre à atmosphère contrôlée	6.3.1-**10**
chilling damage	冷却损坏	altération due à la réfrigération	6.3.1-**11**
climacteric rise	跃变临界期	crise climactérique	6.3.1-**12**
controlled-atmosphere storage *C.A. storage* ◐ *storage in controlled atmosphere* ◐	气调贮藏	entreposage en atmosphère contrôlée (AC) *entreposage AC*	6.3.1-**13**
controlled-atmosphere (storage) room *C.A. storage room* ◐ *gas store* ○	气调贮藏室	chambre à atmosphère contrôlée	6.3.1-**14**
core flush	果心变褐	cœur rosé	6.3.1-**15**
cultivar *variety* ◐	栽培品种	cultivar *variété*	6.3.1-**16**
damage	损坏	dommage *dégât* ◐	6.3.1-**17**
degree of maturity	成熟度	degré de maturité	6.3.1-**18**
degreening	褪绿	déverdissage	6.3.1-**19**
exchanger-diffuser	气体交换扩散器	échangeur-diffuseur	6.3.1-**20**
fruit packing station	水果收购站	station fruitière *station de conditionnement* ◐	6.3.1-**21**
fruit and vegetables in bulk	散装果蔬	fruits et légumes en vrac	6.3.1-**22**
fruit and vegetables in layers	果蔬分层存放	fruits et légumes lités	6.3.1-**23**
gas store ○ *controlled-atmosphere (storage) room* *C.A. storage room* ◐	气调贮藏室	chambre à atmosphère contrôlée	6.3.1-**24**
grading *size grading*	分级 规格分级	calibrage	6.3.1-**25**
growth defect	生长缺陷	défaut de croissance *défaut de développement* ◐	6.3.1-**26**
immature	不成熟的	immature	6.3.1-**27**
infrared CO_2-meter	红外线二氧化碳测定仪	analyseur de CO_2 par infra-rouge *détecteur par infra-rouge* ◐	6.3.1-**28**
internal breakdown *low-temperature breakdown* ◐	低温破坏	brunissement interne *décomposition interne* ◐ *dégradation interne* ◐	6.3.1-**29**
internal defect	内在缺陷	défaut interne	6.3.1-**30**
low-temperature breakdown ◐ *internal breakdown*	低温破坏	brunissement interne *décomposition interne* ◐ *dégradation interne* ◐	6.3.1-**31**
nitrogen flushing	充氮	balayage à l'azote	6.3.1-**32**

	ENGLISH	汉语	FRANÇAIS
6.3.1-**33**	Orsat apparatus	奥氏气体分析仪	appareil d'Orsat
6.3.1-**34**	over-mature (vegetable)	（蔬菜的）过度成熟	montée (d'un légume)
6.3.1-**35**	over-ripeness (of a fruit)	（水果的）过度成熟	blettissement (d'un fruit) *surmaturité* ◐
6.3.1-**36**	oxygen pull-down	降氧	abaissement du taux d'oxygène *réduction de la teneur en oxygène*
6.3.1-**37**	paramagnetic O_2-meter	顺磁性氧气测定仪	analyseur d'oxygène paramagnétique
6.3.1-**38**	physiological disorder	生理失调	maladie physiologique *trouble physiologique*
6.3.1-**39**	physiological maturity (of a fruit)	（水果）生理成熟度	maturité physiologique (d'un fruit)
6.3.1-**40**	refrigeration section	冷却工段	bloc frigorifique
6.3.1-**41**	regenerative scrubber	再生气体洗涤器	absorbeur de CO_2 régénérable
6.3.1-**42**	ripening	催熟	maturation
6.3.1-**43**	rot	腐烂	pourriture
6.3.1-**44**	russeting	变黄褐色	roussissement *rugosité* ◐
6.3.1-**45**	scald	褐烫	échaudure
6.3.1-**46**	scrubber	气体洗涤器	absorbeur de CO_2
6.3.1-**47**	size grading *grading*	分级 规格分级	calibrage
6.3.1-**48**	storage in controlled atmosphere ◐ *controlled-atmosphere storage* *C.A. storage* ◐	气调贮藏	entreposage en atmosphère contrôlée (AC) *entreposage AC*
6.3.1-**49**	variety ◐ *cultivar*	栽培品种	cultivar *variété*
6.3.1-**50**	vegetable packing station	蔬菜收购站	station légumière

	SECTION 6.3.2 *Meat*	节 **6.3.2** 肉品	**SOUS-CHAPITRE 6.3.2** *Produits carnés*
6.3.2-**1**	abattoir *slaughterhouse* ○	屠宰场	abattoir
6.3.2-**2**	ageing (of meat) *maturation*	（肉的）成熟 肉的成熟	maturation (de la viande)
6.3.2-**3**	boned meat	剔骨肉	viande désossée
6.3.2-**4**	bone taint	胴内腐坏	altération en profondeur
6.3.2-**5**	brining	盐浸	saumurage
6.3.2-**6**	carcass	胴体	carcasse
6.3.2-**7**	(carcass) chilling-process	（胴体）冷却工序	réfrigération primaire (des carcasses) *ressuage réfrigéré* ◐
6.3.2-**8**	cold shortening	冷缩	contraction par le froid *raccourcissement dû au froid* ◐
6.3.2-**9**	curing cellar	腌制室	chambre de salaison
6.3.2-**10**	eviscerated poultry	去内脏家禽	volaille éviscérée
6.3.2-**11**	gutted poultry	去肠家禽	volaille effilée

ENGLISH	汉语	FRANÇAIS	
hot deboning	常温剔骨	désossage à chaud	6.3.2-**12**
maturation *ageing (of meat)*	（肉的）成熟 肉的成熟	maturation (de la viande)	6.3.2-**13**
(meat) ageing room ◐	（肉）成熟间	chambre de maturation (de la viande)	6.3.2-**14**
(meat) cutting room	（肉）分割间	salle de découpe (de viande)	6.3.2-**15**
meat holding room	存肉间	chambre de conservation (de viande) *resserre à viande* ◐	6.3.2-**16**
offal	下水	abats	6.3.2-**17**
quarter (of meat)	（牛或马）四分之一胴体	quartier (de viande)	6.3.2-**18**
sausage drying room	香肠干燥间	séchoir à saucisson	6.3.2-**19**
shock-chilling	急速冷却	réfrigération-choc	6.3.2-**20**
slaughterhouse ○ *abattoir*	屠宰场	abattoir	6.3.2-**21**
smoking	烟熏	fumaison	6.3.2-**22**
spin chiller	旋转冷却器	refroidisseur à tambour-agitateur	6.3.2-**23**
thaw rigor	解冻僵直	rigor à la décongélation *rigidité à la décongélation* ◐	6.3.2-**24**
tripe	肚	tripes	6.3.2-**25**

SECTION 6.3.3 *Dairy products*	节 **6.3.3** 奶制品	**SOUS-CHAPITRE 6.3.3** *Produits laitiers*	
bulk collection of milk	牛奶的散装收集	ramassage du lait en vrac	6.3.3-**1**
bulk milk cooler (direct-expansion type)	直接膨胀式散装牛奶冷却器	refroidisseur de lait en vrac à détente directe	6.3.3-**2**
bulk milk cooler (iced-water type)	冰水式散装牛奶冷却器	refroidisseur de lait en vrac à eau glacée	6.3.3-**3**
butter factory	黄油工厂	beurrerie	6.3.3-**4**
butter making	黄油制作	beurrerie	6.3.3-**5**
can milk cooler (USA) *churn milk cooler (UK)*	搅拌式牛奶冷却器	refroidisseur de lait en bidons *refroidisseur de lait en pots* ◐	6.3.3-**6**
can sprinkling cooler *cascade milk cooler* *irrigation cooler*	喷淋式搅奶冷却器	refroidisseur de lait à aspersion	6.3.3-**7**
cascade milk cooler *can sprinkling cooler* *irrigation cooler*	喷淋式搅奶冷却器	refroidisseur de lait à aspersion	6.3.3-**8**
central milk plant	中心牛奶工厂	centrale laitière	6.3.3-**9**
cheese curing *cheese ripening*	奶酪成熟	affinage du fromage	6.3.3-**10**
cheese drying	奶酪干燥	hâlage du fromage	6.3.3-**11**
cheese drying room	奶酪干燥室	hâloir à fromage	6.3.3-**12**
cheese factory	奶酪工厂	fromagerie	6.3.3-**13**
cheese making	奶酪制作	fromagerie	6.3.3-**14**
cheese ripening *cheese curing*	奶酪成熟	affinage du fromage	6.3.3-**15**

	ENGLISH	汉语	FRANÇAIS
6.3.3-**16**	churn immersion cooler *immersion milk cooler*	沉浸式搅奶冷却器	refroidisseur de bidons de lait à immersion
6.3.3-**17**	churn milk cooler (UK) *can milk cooler (USA)*	搅拌式牛奶冷却器	refroidisseur de lait en bidons *refroidisseur de lait en pots* ◐
6.3.3-**18**	churning	搅奶	barattage
6.3.3-**19**	city milk plant	城市牛奶工厂	centrale laitière
6.3.3-**20**	clotting *coagulation*	凝结	caillage *coagulation*
6.3.3-**21**	coagulation *clotting*	凝结	caillage *coagulation*
6.3.3-**22**	curd	凝乳 凝固奶酪	caillé
6.3.3-**23**	dairy factory *milk plant* *dairy plant* *milk factory*	乳品工厂 乳品加工厂	laiterie (usine) *usine laitière*
6.3.3-**24**	dairy plant *milk plant* *dairy factory* *milk factory*	乳品工厂 乳品加工厂	laiterie (usine) *usine laitière*
6.3.3-**25**	dairy shop	乳品商店	crémerie *laiterie (boutique)* ◐
6.3.3-**26**	immersion milk cooler *churn immersion cooler*	沉浸式搅奶冷却器	refroidisseur de bidons de lait à immersion
6.3.3-**27**	in-can immersion cooler *plunger-type milk cooler*	插入式牛奶冷却器	refroidisseur de lait à plongeur
6.3.3-**28**	irrigation cooler *can sprinkling cooler* *cascade milk cooler*	喷淋式搅奶冷却器	refroidisseur de lait à aspersion
6.3.3-**29**	milk collection centre	牛奶集中站	centre de ramassage de lait *centre de collecte de lait*
6.3.3-**30**	milk collection in cans (USA) *milk collection in churns (UK)*	搅奶器收集	ramassage du lait en bidons *ramassage du lait en pots* ◐
6.3.3-**31**	milk collection in churns (UK) *milk collection in cans (USA)*	搅奶器收集	ramassage du lait en bidons *ramassage du lait en pots* ◐
6.3.3-**32**	milk cooler	牛奶冷却器	refroidisseur de lait
6.3.3-**33**	milk cooling on the farm	牛奶农场冷却	refroidissement du lait à la ferme
6.3.3-**34**	milk factory *milk plant* *dairy factory* *dairy plant*	乳品工厂 乳品加工厂	laiterie (usine) *usine laitière*
6.3.3-**35**	milk plant *dairy factory* *dairy plant* *milk factory*	乳品工厂 乳品加工厂	laiterie (usine) *usine laitière*
6.3.3-**36**	plunger-type milk cooler *in-can immersion cooler*	插入式牛奶冷却器	refroidisseur de lait à plongeur
6.3.3-**37**	refrigerated farm tank	农场冷却罐	bac refroidisseur de ferme *cuve de refroidissement à la ferme* ◐
6.3.3-**38**	sparge-ring-type milk cooler ○	喷淋环式搅奶冷却器	refroidisseur de lait à collier d'aspersion
6.3.3-**39**	turbine milk cooler	涡轮牛奶冷却器	refroidisseur de lait à tourniquet
6.3.3-**40**	uperization ○	高温消毒	upérisation

ENGLISH	汉语	FRANÇAIS	
SECTION 6.3.4 *Ice cream*	节 **6.3.4** 冰淇淋	**SOUS-CHAPITRE 6.3.4** *Crèmes glacées*	
batch-type (ice cream) freezer	间歇式（冰淇淋）冻结器	congélateur discontinu (de crème glacée) *turbine à crème glacée* ◐	6.3.4-**1**
conservator ○ *ice cream cabinet*	冰淇淋柜	conservateur de crème glacée	6.3.4-**2**
contact hardening	接触硬化	durcissement par contact	6.3.4-**3**
continuous (ice cream) freezer	连续式（冰淇淋）冻结器	congélateur continu (de crème glacée)	6.3.4-**4**
hardening	硬化	durcissement	6.3.4-**5**
ice cream	冰淇淋	crème glacée *glace* ◐	6.3.4-**6**
ice cream brick	冰砖	pavé de glace	6.3.4-**7**
ice cream cabinet *conservator* ○	冰淇淋柜	conservateur de crème glacée	6.3.4-**8**
ice cream freezer	冰淇淋冻结器	congélateur de crème glacée	6.3.4-**9**
(ice cream) hardening cabinet	（冰淇淋）硬化柜	meuble de durcissement (de crème glacée)	6.3.4-**10**
(ice cream) hardening room	（冰淇淋）硬化室	chambre de durcissement (de crème glacée)	6.3.4-**11**
ice cream manufacturer	冰淇淋制造商	glacier	6.3.4-**12**
(ice cream) mix	（冰淇淋）配料	mélange (pour crème glacée) *mix* ◐	6.3.4-**13**
ice lolly	冰棍 冰糕	bâtonnet glacé	6.3.4-**14**
ice milk	冰奶	glace au lait	6.3.4-**15**
over-run	泡起现象	foisonnement	6.3.4-**16**
rapid hardener	快速硬化器	installation de durcissement rapide	6.3.4-**17**
sherbet (USA) *sorbet* *water ice* ◐	果汁冰块	sorbet	6.3.4-**18**
soft ice (cream)	软冰淇淋	glace "à l'italienne" *glace molle* ◐	6.3.4-**19**
sorbet *sherbet (USA)* *water ice* ◐	果汁冰块	sorbet	6.3.4-**20**
tunnel hardening	隧道硬化	durcissement en tunnel	6.3.4-**21**
water ice ◐ *sorbet* *sherbet (USA)*	果汁冰块	sorbet	6.3.4-**22**

ENGLISH	汉语	FRANÇAIS	
SECTION 6.3.5 *Fish and seafood*	节 **6.3.5** 鱼与海产品	**SOUS-CHAPITRE 6.3.5** *Produits de la pêche*	
autolysis (of fish)	（鱼）自变质	autolyse	6.3.5-**1**
bilgy fish ○ *stinker* ○	臭鱼	poisson fangeux ○	6.3.5-**2**
chilled sea water (CSW)	冰-（盐）水混合物 （CSW）	eau de mer fraîche	6.3.5-**3**

	ENGLISH	汉语	FRANÇAIS
6.3.5-4	filleting (of fish)	切鱼片	filetage (du poisson)
6.3.5-5	icing	加冰冷藏	glaçage
6.3.5-6	refrigerated sea water (RSW)	冷藏用海水 （RSW）	eau de mer réfrigérée
6.3.5-7	rust	锈色	rouille
6.3.5-8	stinker ○ *bilgy fish* ○	臭鱼	poisson fangeux ○

	SECTION 6.3.6 *Beverages*	节 **6.3.6** 饮料	**SOUS-CHAPITRE 6.3.6** *Boissons*
6.3.6-1	beer cooler	啤酒冷却器	refroidisseur de bière
6.3.6-2	bottom fermentation *low fermentation* ◐	底层发酵	fermentation basse
6.3.6-3	carbonator	碳酸化器	saturateur (en CO_2)
6.3.6-4	chill proofing *cold stabilization* ◐	冷稳定	stabilisation par le froid
6.3.6-5	cold stabilization ◐ *chill proofing*	冷稳定	stabilisation par le froid
6.3.6-6	fermenting cellar	发酵地窖	salle de fermentation
6.3.6-7	high fermentation ◐ *top fermentation*	上层发酵	fermentation haute
6.3.6-8	low fermentation ◐ *bottom fermentation*	底层发酵	fermentation basse
6.3.6-9	stock cellar	贮酒地窖	cave de garde
6.3.6-10	top fermentation *high fermentation* ◐	上层发酵	fermentation haute
6.3.6-11	wine factory ○ *winery*	葡萄酒厂	cave de vinification
6.3.6-12	winery *wine factory* ○	葡萄酒厂	cave de vinification
6.3.6-13	wort cooler	麦芽汁冷却器	refroidisseur de moût

	SECTION 6.3.7 *Other products* *(egg products, sweets, etc.)*	节 **6.3.7** 其它产品（蛋制品，糖果等）	**SOUS-CHAPITRE 6.3.7** *Autres produits* *(ovoproduits, confiserie, etc.)*
6.3.7-1	bakery (refrigerated) slab	揉面（冷的）石板	table froide *tour de pâtisserie* ◐
6.3.7-2	bottomer slab	打底冷板	plaque de durcissement des fonds
6.3.7-3	egg-breaking plant (USA) ◐ *egg-shelling plant*	打蛋厂	casserie d'œufs
6.3.7-4	egg candling	照蛋	mirage des œufs
6.3.7-5	egg-shelling plant *egg-breaking plant (USA)* ◐	打蛋厂	casserie d'œufs

ENGLISH	汉语	FRANÇAIS	
sponge dough	发酵面团	pâte pain-levure *pâte "semi-directe"* ◐	6.3.7-**6**
staling	陈旧	rassissement	6.3.7-**7**
straight dough	未发酵面团	pâte directe	6.3.7-**8**
winterisation	冻凝	frigélisation	6.3.7-**9**

SECTION 6.4 *Other treatments used in the food industry*	节 **6.4** 食品工业中所用的其它处理方法	**SOUS-CHAPITRE 6.4** *Autres traitements dans l'industrie agroalimentaire*	
adjunct to refrigeration *supplement to refrigeration* ◐	冷冻辅助手段	adjuvant du froid	6.4-**1**
alginate coating	涂藻酸盐	revêtement gélifiant	6.4-**2**
anti-oxidant	抗氧化剂	anti-oxydant *anti-oxygène* ◐	6.4-**3**
antibiotic	抗生素	antibiotique	6.4-**4**
appertization	食品的密封杀菌保存	appertisation	6.4-**5**
bactericide	杀菌剂	bactéricide	6.4-**6**
bacteriostatic	抑菌剂	bactériostatique	6.4-**7**
blanching	烫漂	blanchiment	6.4-**8**
camouflage of goods	商品的伪装	fardage	6.4-**9**
canned food (USA) *tinned food* ◐	罐头食品	conserve appertisée	6.4-**10**
food additive	食品添加剂	additif alimentaire	6.4-**11**
fumigation	薰蒸	fumigation	6.4-**12**
fungicide	杀真菌剂	fongicide	6.4-**13**
gel *jelly* ◐	凝胶	gel *gelée* ◐	6.4-**14**
gelation	胶化	gélification	6.4-**15**
ionizing radiation	电离辐射	rayonnements ionisants	6.4-**16**
irradiation	辐照	irradiation	6.4-**17**
jelly ◐ *gel*	凝胶	gel *gelée* ◐	6.4-**18**
odour control	气味控制	lutte contre les odeurs	6.4-**19**
oiling	涂油	huilage	6.4-**20**
pasteurization	巴氏杀菌	pasteurisation	6.4-**21**
preheating	预热	préchauffage	6.4-**22**
preservative	防腐剂	agent conservateur *conservateur*	6.4-**23**
radiation pasteurization *radiopasteurization* ◐ *radurization* ◐	辐照杀菌	radiopasteurisation	6.4-**24**
radiation sterilization *radioappertization* ◐ *radiosterilization* ◐	辐照消毒	radiostérilisation *radappertisation* ○ *radioappertisation* ○	6.4-**25**

	ENGLISH	汉语	FRANÇAIS
6.4-**26**	radicidation	低剂量射线处理	radicidation
6.4-**27**	radioappertization ◐ *radiation sterilization* *radiosterilization* ◐	辐照消毒	radiostérilisation *radappertisation* ○ *radioappertisation* ○
6.4-**28**	radiopasteurization ◐ *radiation pasteurization* *radurization* ◐	辐照杀菌	radiopasteurisation
6.4-**29**	radiosterilization ◐ *radiation sterilization* *radioappertization* ◐	辐照消毒	radiostérilisation *radappertisation* ○ *radioappertisation* ○
6.4-**30**	radurization ◐ *radiation pasteurization* *radiopasteurization* ◐	辐照杀菌	radiopasteurisation
6.4-**31**	scalding	热烫	échaudage
6.4-**32**	semi-preserve	半保存	semi-conserve
6.4-**33**	steaming	蒸汽处理	ébouillantage
6.4-**34**	sterilization	消毒	stérilisation
6.4-**35**	supplement to refrigeration ◐ *adjunct to refrigeration*	冷冻辅助手段	adjuvant du froid
6.4-**36**	tinned food ◐ *canned food (USA)*	罐头食品	conserve appertisée
6.4-**37**	warming room	加热（升温）间 升温间	salle de réchauffage
6.4-**38**	wax coating	涂腊	revêtement cireux *revêtement paraffineux* ◐

	SECTION 6.5 *Packaging*	节 **6.5** 包装	**SOUS-CHAPITRE 6.5** *Conditionnement et emballage*
6.5-**1**	active packaging	主动包装 活性包装	emballage actif *emballage en atmosphère modifiée* ◐
6.5-**2**	batch *lot*	一批	lot
6.5-**3**	lot *batch*	一批	lot
6.5-**4**	marking	商标 标记	marquage
6.5-**5**	modified atmosphere packaging (MAP)	气调包装（MAP）	emballage sous atmosphère modifiée
6.5-**6**	packaging *packing*	包装 打包 装箱	conditionnement *emballage*
6.5-**7**	packing *packaging*	包装 打包 装箱	conditionnement *emballage*
6.5-**8**	packing station	包装站	centre de conditionnement *station de conditionnement* *plate-forme de conditionnement* ◐ *station d'emballage* ◐
6.5-**9**	prepackaging ◐ *prepacking*	预包装	préemballage

ENGLISH	汉语	FRANÇAIS	
prepacking *prepackaging* ◐	预包装	préemballage	6.5-**10**
sorting	分类	triage	6.5-**11**
transport packaging	运输包装	suremballage	6.5-**12**
ungrouping	分批 分类（发货）	dégroupage	6.5-**13**
vacuum packing	真空包装	conditionnement sous vide *emballage sous vide* ◐	6.5-**14**

章 7. 空气调节

◐ 容许（或默许）术语

○ 过时术语

ENGLISH	汉语	FRANÇAIS	
SECTION 7.1 *Air conditioning: general background*	节 **7.1** 空气调节：基础知识	**SOUS-CHAPITRE 7.1** *Généralités sur le conditionnement d'air*	
acclimation (USA) *acclimatization*	环境适应性	acclimatation	7.1-**1**
acclimatization *acclimation (USA)*	环境适应性	acclimatation	7.1-**2**
air conditioning (AC)	空气调节	conditionnement d'air	7.1-**3**
air-conditioning installation *air-conditioning plant*	空调机组	installation de conditionnement d'air	7.1-**4**
air-conditioning plant *air-conditioning installation*	空调机组	installation de conditionnement d'air	7.1-**5**
air-conditioning process *air-conditioning system*	空调过程	procédé de conditionnement d'air *système de conditionnement d'air*	7.1-**6**
air-conditioning system *air-conditioning process*	空调过程	procédé de conditionnement d'air *système de conditionnement d'air*	7.1-**7**
air handling *air treatment*	空气处理	traitement de l'air	7.1-**8**
air treatment *air handling*	空气处理	traitement de l'air	7.1-**9**
all-year air conditioning *year-round air conditioning*	全年性空气调节	conditionnement d'air toutes saisons *conditionnement d'air été-hiver* ◐	7.1-**10**
ambient air	环境空气	air ambiant	7.1-**11**
apparatus dew point	仪器露点	point de rosée équivalent	7.1-**12**
atmosphere of reference ◐	基准大气 参考大气条件	atmosphère de référence ◐	7.1-**13**
chiller	冷水机组	groupe refroidisseur d'eau	7.1-**14**
climatic engineering ○ *environmental engineering* ◐	环境工程	génie climatique	7.1-**15**
comfort air conditioning	舒适性空调	climatisation *conditionnement d'air de confort* ◐	7.1-**16**
comfort cooling	舒适性降温	rafraîchissement pour le confort	7.1-**17**
comfort index	舒适性指标	indice de confort	7.1-**18**
conditioned air	调节过的空气	air conditionné	7.1-**19**
corrected effective temperature	校正的有效温度	température effective corrigée	7.1-**20**
degree-day *kelvin-day* ◐	度日	degré-jour	7.1-**21**
dehumidifying effect	除湿效应	puissance de déshumidification *puissance frigorifique latente (d'un refroidisseur d'air)* ◐	7.1-**22**
design conditions	设计条件 设计工况	conditions contractuelles	7.1-**23**
effective temperature	有效温度	température effective	7.1-**24**
environment cooling ◐	环境冷却	refroidissement d'une ambiance	7.1-**25**
environmental conditions	环境条件	conditions d'ambiance	7.1-**26**
environmental engineering ◐ *climatic engineering* ○	环境工程	génie climatique	7.1-**27**

	ENGLISH	汉语	FRANÇAIS
7.1-**28**	equivalent temperature	当量温度	température équivalente *température résultante sèche*
7.1-**29**	eupatheoscope	舒适计	eupathéoscope
7.1-**30**	freshness index	新鲜度指标	indice de fraîcheur
7.1-**31**	frigorimeter	低温计	frigorimètre
7.1-**32**	globe thermometer	套球式温度计	thermomètre globe
7.1-**33**	greenhouse effect	温室效应	effet de serre
7.1-**34**	humidifying effect	加湿效果	puissance d'humidification
7.1-**35**	indoor climate	室内气候	climat intérieur *climat confiné*
7.1-**36**	indoor conditions	室内工况 室内条件	conditions intérieures
7.1-**37**	industrial air conditioning *process air conditioning* ◐	工业空调	conditionnement d'air industriel
7.1-**38**	insolation	日射率 日照	insolation
7.1-**39**	katathermometer	卡他温度计	catathermomètre
7.1-**40**	kelvin-day ◐ *degree-day*	度日	degré-jour
7.1-**41**	latent cooling capacity	潜热制冷量	puissance frigorifique latente
7.1-**42**	leakage airflow	漏风量	fuite d'air
7.1-**43**	low-grade heat source	低品位热源 低热能源	source de chaleur à basse température
7.1-**44**	marine air conditioning	船舶空调	conditionnement d'air à bord des navires
7.1-**45**	microclimate	微气候	microclimat
7.1-**46**	multizone	多区域	multizone
7.1-**47**	net cooling capacity	净冷却能力 净制冷量（总冷量－风机耗能）	puissance frigorifique nette
7.1-**48**	net total cooling capacity	总冷却能力 总制冷量	puissance frigorifique totale nette
7.1-**49**	occupied zone	占用区	zone occupée
7.1-**50**	outdoor conditions	室外工况 室外条件	conditions extérieures
7.1-**51**	process air conditioning ◐ *industrial air conditioning*	工业空调	conditionnement d'air industriel
7.1-**52**	residential air conditioning	住宅空气调节	conditionnement d'air résidentiel
7.1-**53**	resulting temperature	综合温度	température résultante
7.1-**54**	sensible cooling capacity *sensible (dry) air-cooling capacity* *sensible cooling effect* ○	显热制冷量 显热冷却效应	puissance frigorifique sensible
7.1-**55**	sensible (dry) air-cooling capacity *sensible cooling capacity* *sensible cooling effect* ○	显热制冷量 显热冷却效应	puissance frigorifique sensible
7.1-**56**	sensible cooling effect ○ *sensible cooling capacity* *sensible (dry) air-cooling capacity*	显热制冷量 显热冷却效应	puissance frigorifique sensible

ENGLISH	汉语	FRANÇAIS	
sensible heat ratio	显热比	coefficient de chaleur sensible (d'un refroidisseur d'air)	7.1-**57**
solar heat gain	太阳得热 日照得热	apport de chaleur par insolation	7.1-**58**
standard air	标准大气（空气参数）	air normal	7.1-**59**
standard atmosphere of reference	标准参考大气条件	atmosphère normale de référence	7.1-**60**
total cooling capacity *total cooling effect* ○	总制冷量 总冷却能力 总冷却效果	puissance frigorifique totale	7.1-**61**
total cooling effect ○ *total cooling capacity*	总制冷量 总冷却能力 总冷却效果	puissance frigorifique totale	7.1-**62**
treated air	处理过的空气	air traité	7.1-**63**
year-round air conditioning *all-year air conditioning*	全年性空气调节	conditionnement d'air toutes saisons *conditionnement d'air été-hiver* ◐	7.1-**64**
zoning	分区	zonage	7.1-**65**

SECTION 7.2 *Specific definitions* **SECTION 7.2.1** *Air-conditioning production*	节 **7.2** 专门定义 节 **7.2.1** 空气调节	**SOUS-CHAPITRE 7.2** *Notions spécifiques* **SOUS-CHAPITRE 7.2.1** *Production d'air conditionné*	
air cooler *cooling unit* *unit cooler*	空气冷却器	refroidisseur d'air *frigorifère* ◐	7.2.1-**1**
bypassed indoor airflow	旁通的室内气流	air intérieur recyclé	7.2.1-**2**
bypassed outdoor airflow	旁通的室外气流	air extérieur recyclé	7.2.1-**3**
cooling unit *air cooler* *unit cooler*	空气冷却器	refroidisseur d'air *frigorifère* ◐	7.2.1-**4**
desiccant cooling *desiccant evaporative cooling*	干燥剂冷却	système déshydratant *système déshydratant à évaporation*	7.2.1-**5**
desiccant evaporative cooling *desiccant cooling*	干燥剂冷却	système déshydratant *système déshydratant à évaporation*	7.2.1-**6**
double-spacing finned cooler ◐ *two-way finned cooler*	双片距冷却器	refroidisseur à double écartement d'ailettes	7.2.1-**7**
draught tower	通风塔	tour à vent	7.2.1-**8**
dry cooling coil *dry-surface coil*	干式冷却盘管	batterie sèche	7.2.1-**9**
dry-surface coil *dry cooling coil*	干式冷却盘管	batterie sèche	7.2.1-**10**
dry-type air cooler	干式空气冷却器	refroidisseur d'air du type sec *frigorifère sec* ◐	7.2.1-**11**
forced-circulation air cooler *forced-convection air cooler* ◐ *forced-draught air cooler* ◐	强制循环空气冷却器	refroidisseur d'air à convection forcée *aérofrigorifère* ◐	7.2.1-**12**
forced-convection air cooler ◐ *forced-circulation air cooler* *forced-draught air cooler* ◐	强制循环空气冷却器	refroidisseur d'air à convection forcée *aérofrigorifère* ◐	7.2.1-**13**

	ENGLISH	汉语	FRANÇAIS
7.2.1-**14**	forced-convection air heater ○ *unit heater*	暖风机	aérotherme *réchauffeur d'air à convection forcée*
7.2.1-**15**	forced-draught air cooler ◐ *forced-circulation air cooler* *forced-convection air cooler* ◐	强制循环空气冷却器	refroidisseur d'air à convection forcée *aérofrigorifère* ◐
7.2.1-**16**	gas-fired heater	燃气加热器	dispositif de chauffage au gaz
7.2.1-**17**	heater battery	加热器组	batterie de chauffe
7.2.1-**18**	heater coil *heating resistance* ◐	电加热器	résistance chauffante
7.2.1-**19**	heating coil	加热盘管	serpentin de chauffage *tube chauffant*
7.2.1-**20**	heating resistance ◐ *heater coil*	电加热器	résistance chauffante
7.2.1-**21**	indoor discharge airflow	室内空气排放量	air intérieur soufflé
7.2.1-**22**	indoor heat exchanger	室内侧热交换器	échangeur thermique intérieur
7.2.1-**23**	indoor intake airflow	室内回风循环量	air intérieur repris
7.2.1-**24**	natural-convection air cooler	自然对流空气冷却器	refroidisseur d'air à convection naturelle
7.2.1-**25**	non-ducted indoor air-conditioning equipment	无风管室内空气调节机（组）	climatiseur intérieur non raccordé
7.2.1-**26**	outdoor-discharge airflow	室外机排风量	air extérieur refoulé
7.2.1-**27**	outdoor heat exchanger	室外侧热交换器	échangeur thermique extérieur
7.2.1-**28**	outdoor intake airflow	室外机进风量	air extérieur aspiré
7.2.1-**29**	panel cooler	平板冷却器	panneau refroidisseur
7.2.1-**30**	preheating	预热	préchauffage
7.2.1-**31**	radiant cooling system	辐射冷却系统	système de refroidissement par panneaux
7.2.1-**32**	reheating	再加热	réchauffage
7.2.1-**33**	self-contained air cooler *self-contained cooling unit*	整体式空气冷却机组	refroidisseur d'air à groupe incorporé
7.2.1-**34**	self-contained cooling unit *self-contained air cooler*	整体式空气冷却机组	refroidisseur d'air à groupe incorporé
7.2.1-**35**	sensible-heat air cooler	显热空气冷却器	refroidisseur d'air à chaleur sensible
7.2.1-**36**	spray-type air cooler	喷淋式空气冷却器	refroidisseur d'air à pulvérisation *frigorifère à pulvérisation* ◐
7.2.1-**37**	troffer ◐ *ventilated light fitting*	通风照明装置	luminaire ventilé
7.2.1-**38**	two-way finned cooler *double spacing finned cooler* ◐	双片距冷却器	refroidisseur à double écartement d'ailettes
7.2.1-**39**	unit cooler *air cooler* *cooling unit*	空气冷却器	refroidisseur d'air *frigorifère* ◐
7.2.1-**40**	unit heater *forced-convection air heater* ○	暖风机	aérotherme *réchauffeur d'air à convection forcée*
7.2.1-**41**	ventilated light fitting *troffer* ◐	通风照明装置	luminaire ventilé
7.2.1-**42**	wet-type air cooler	湿式空气冷却器	refroidisseur d'air du type humide *frigorifère humide* ◐

ENGLISH	汉语	FRANÇAIS	
SECTION 7.2.2 *Air circulation and distribution*	节 **7.2.2** 空气循环及输配	**SOUS-CHAPITRE 7.2.2** *Circulation et distribution de l'air*	
aeration	通风	aération	7.2.2-**1**
air change *ventilation rate*	换气 换气次数	renouvellement d'air	7.2.2-**2**
air circulation	空气循环	circulation d'air	7.2.2-**3**
air-circulation ratio *rate of air circulation*	室内空气循环率	coefficient de brassage *taux de brassage*	7.2.2-**4**
air diffusion	空气扩散	diffusion d'air	7.2.2-**5**
air distribution	空气输配	distribution d'air	7.2.2-**6**
(air) exfiltration	(空气)漏出	exfiltration (d'air)	7.2.2-**7**
air exhaust *air extract*	排气	évacuation d'air	7.2.2-**8**
air extract *air exhaust*	排气	évacuation d'air	7.2.2-**9**
(air) infiltration	(空气)渗入	infiltration (d'air)	7.2.2-**10**
air inlet *air intake*	进气口	entrée d'air (neuf) *prise d'air (neuf)*	7.2.2-**11**
air intake *air inlet*	进气口	entrée d'air (neuf) *prise d'air (neuf)*	7.2.2-**12**
air velocity	空气流速	vitesse d'air	7.2.2-**13**
airflow resistance	气流阻力	résistance à l'écoulement de l'air	7.2.2-**14**
airing	自然通风	aération	7.2.2-**15**
attic ventilation	阁楼通风	ventilation des combles	7.2.2-**16**
axial velocity	轴向速度	vitesse axiale	7.2.2-**17**
blow ◐ *throw*	射程	portée	7.2.2-**18**
chimney effect *stack effect ◐*	烟囱效应	tirage	7.2.2-**19**
Coanda effect	附壁效应 康达效应	effet Coanda	7.2.2-**20**
cooling air	冷却空气	air de refroidissement	7.2.2-**21**
cross ventilation	贯流通风	ventilation transversale	7.2.2-**22**
dead air pocket ◐ *dead zone ○*	滞流区	zone morte	7.2.2-**23**
dead zone ○ *dead air pocket ◐*	滞流区	zone morte	7.2.2-**24**
delivery air *supply air*	送风	air fourni *air soufflé*	7.2.2-**25**
diffusion area	扩散面积	surface balayée *surface ventilée ◐*	7.2.2-**26**
displacement air diffusion	置换式空气扩散	diffusion d'air par déplacement	7.2.2-**27**
draft (USA) *draught*	引风 (压差)通风	courant d'air *appel d'air*	7.2.2-**28**

	ENGLISH	汉语	FRANÇAIS
7.2.2-**29**	draught *draft (USA)*	引风 （压差）通风	courant d'air *appel d'air*
7.2.2-**30**	drop	（气流）落差	retombée
7.2.2-**31**	entrainment ◐ *induction*	诱导	induction
7.2.2-**32**	entrainment ratio ◐ *induction ratio*	诱导比	taux d'induction *coefficient d'induction* ◐
7.2.2-**33**	exhaust air *extracted air* ◐	排风	air évacué *air extrait*
7.2.2-**34**	exhaust airflow	排放空气流	air rejeté
7.2.2-**35**	exit air	排风	air rejeté
7.2.2-**36**	extracted air ◐ *exhaust air*	排风	air évacué *air extrait*
7.2.2-**37**	face velocity *frontal velocity* ◐	迎面风速	vitesse frontale
7.2.2-**38**	forced-air circulation	强制空气循环	ventilation forcée *circulation d'air forcée*
7.2.2-**39**	(fresh) air make-up	新风补充量	air d'appoint
7.2.2-**40**	fresh air *outdoor air* *replacement air* *outside air* ◐	新风 室外空气 新鲜空气	air hygiénique *air neuf* *air extérieur* ◐ *air frais* ◐
7.2.2-**41**	frontal velocity ◐ *face velocity*	迎面风速	vitesse frontale
7.2.2-**42**	induction *entrainment* ◐	诱导	induction
7.2.2-**43**	induction ratio *entrainment ratio* ◐	诱导比	taux d'induction *coefficient d'induction* ◐
7.2.2-**44**	mixing air diffusion	混合式空气扩散	diffusion d'air par mélange
7.2.2-**45**	natural air circulation	自然空气循环	circulation d'air naturelle
7.2.2-**46**	non-isothermal jet	非等温射流	jet non isotherme
7.2.2-**47**	outdoor air *fresh air* *replacement air* *outside air* ◐	新风 室外空气 新鲜空气	air hygiénique *air neuf* *air extérieur* ◐ *air frais* ◐
7.2.2-**48**	outside air ◐ *fresh air* *outdoor air* *replacement air*	新风 室外空气 新鲜空气	air hygiénique *air neuf* *air extérieur* ◐ *air frais* ◐
7.2.2-**49**	primary air	一次空气	air primaire
7.2.2-**50**	radius of diffusion	扩散半径	rayon de diffusion
7.2.2-**51**	rate of air circulation *air-circulation ratio*	室内空气循环率	coefficient de brassage *taux de brassage*
7.2.2-**52**	recirculated air	再循环空气	air recyclé *air recirculé* ◐
7.2.2-**53**	replacement air ◐ *fresh air* *outdoor air* *outside air* ◐	新风 室外空气 新鲜空气	air hygiénique *air neuf* *air extérieur* ◐ *air frais* ◐
7.2.2-**54**	return air	回风	air repris

ENGLISH	汉语	FRANÇAIS	
secondary air	二次空气	air secondaire	7.2.2-**55**
spread	射流扩散角	étalement	7.2.2-**56**
stack effect ◐ *chimney effect*	烟囱效应	tirage	7.2.2-**57**
supply air *delivery air*	送风	air fourni *air soufflé*	7.2.2-**58**
terminal velocity	末端速度	vitesse terminale	7.2.2-**59**
throw *blow* ◐	射程	portée	7.2.2-**60**
ventilation	通风	ventilation	7.2.2-**61**
ventilation airflow	通风空气流	air neuf	7.2.2-**62**
ventilation rate *air change*	换气 换气次数	renouvellement d'air	7.2.2-**63**

SECTION 7.2.3 *Air quality*	节 **7.2.3** 空气质量	**SOUS-CHAPITRE 7.2.3** *Qualité de l'air*	
aerosol	气溶胶	aérosol	7.2.3-**1**
air contaminant	大气污染物质	agent de contamination de l'air *contaminant de l'air* *polluant de l'air*	7.2.3-**2**
air pollutant	大气污染	polluant de l'air	7.2.3-**3**
airborne particles *particulates (USA)*	大气尘 微粒	particules en suspension (dans l'air)	7.2.3-**4**
biological agent	生物因子（病原体）	contaminant biologique	7.2.3-**5**
breathing zone	呼吸区	zone respiratoire	7.2.3-**6**
chemical agent	化学介质	contaminant chimique	7.2.3-**7**
decipol	人感觉空气质量的单位	décipol	7.2.3-**8**
deodorization *deodorizing* *odour removal* ◐	除臭	désodorisation	7.2.3-**9**
deodorizing *deodorization* *odour removal* ◐	除臭	désodorisation	7.2.3-**10**
exposure (by inhalation)	接触（遭受）	exposition (par inhalation)	7.2.3-**11**
indoor air quality	室内空气品质	qualité de l'air intérieur	7.2.3-**12**
odour	气味	odeur	7.2.3-**13**
odour removal ◐ *deodorization* *deodorazing*	除臭	désodorisation	7.2.3-**14**
olf	1 olf 为人均气味发生量	olf	7.2.3-**15**
particulates (USA) *airborne particles*	大气尘 微粒	particules en suspension (dans l'air)	7.2.3-**16**
separation efficiency	分离效率	rendement d'un filtre	7.2.3-**17**
smell	臭味	mauvaise odeur	7.2.3-**18**
suspended matter	悬浮粒子	matières en suspension	7.2.3-**19**

	ENGLISH	汉语	FRANÇAIS
	SECTION 7.3 *Specific equipment* **SECTION 7.3.1** *Humidity and temperature control*	节 **7.3** 专用设备 节 **7.3.1** 湿度及温度控制	**SOUS-CHAPITRE 7.3** *Matériels spécifiques* **SOUS-CHAPITRE 7.3.1** *Régulation de l'hygrométrie et de la température*
7.3.1-**1**	activated alumina	活性矾土	alumine activée
7.3.1-**2**	air washer *scrubber*	空气洗涤器	laveur d'air
7.3.1-**3**	atomize (to)	雾化	atomiser
7.3.1-**4**	capillary air washer *capillary humidifier* *cell-type air washer*	蜂窝式填料洗涤器	laveur d'air à alvéoles *humidificateur à alvéoles*
7.3.1-**5**	capillary humidifier *capillary air washer* *cell-type air washer*	蜂窝式填料洗涤器	laveur d'air à alvéoles *humidificateur à alvéoles*
7.3.1-**6**	cell-type air washer *capillary air washer* *capillary humidifier*	蜂窝式填料洗涤器	laveur d'air à alvéoles *humidificateur à alvéoles*
7.3.1-**7**	dehumidification capacity	除湿能力	pouvoir de déshumidification
7.3.1-**8**	dehumidification efficiency ratio (DER)	除湿能效比（DER）	coefficient d'efficacité de déshumidification
7.3.1-**9**	dehumidification for comfort	舒适性除湿	déshumidification pour confort
7.3.1-**10**	dehumidification for process	工艺性除湿	déshumidification pour procédé
7.3.1-**11**	dehumidifier	除湿器	déshumidificateur
7.3.1-**12**	desiccant	干燥剂	déshydratant
7.3.1-**13**	desiccant contactor	除湿段	contacteur du déshydratant
7.3.1-**14**	desiccant wheel	转轮除湿机	roue déshydratante
7.3.1-**15**	drift	漂水	eau entraînée *entraînement vésiculaire* ◐
7.3.1-**16**	eliminator	挡水板	séparateur de gouttelettes *éliminateur de gouttelettes* ◐
7.3.1-**17**	fan-pad system	湿填料蒸发冷却器	refroidisseur d'air à tampon humide *bourrage ventilé*
7.3.1-**18**	humidification rate	加湿率	taux d'humidification
7.3.1-**19**	humidifier	加湿器	humidificateur
7.3.1-**20**	injection (steam) humidifier	(蒸汽)喷射加湿器	humidificateur à injection de vapeur
7.3.1-**21**	liquid desiccant	液体吸湿剂	déshydratant liquide *solution déshydratante*
7.3.1-**22**	liquid desiccant concentration	液态干燥剂浓度	concentration en déshydratant liquide
7.3.1-**23**	liquid desiccant transfer to conditioner	液态干燥剂循环量	transfert du déshydratant liquide au conditionneur
7.3.1-**24**	mean radiant temperature	平均辐射温度	température radiante moyenne
7.3.1-**25**	moisture-removal capacity	除湿量	rendement d'enlèvement de l'humidité
7.3.1-**26**	moisture-removal rate	除湿率	taux d'enlèvement de l'humidité
7.3.1-**27**	molecular sieve	分子筛	tamis moléculaire *crible moléculaire* ◐

ENGLISH	汉语	FRANÇAIS	
pad	填料	milieu dispersant *médium dispersant*	7.3.1-**28**
plane radiant temperature	平面辐射温度	température plane radiante	7.3.1-**29**
process air	待除湿的空气	air à déshumidifier	7.3.1-**30**
regain (of moisture)	再吸湿	reprise (d'humidité)	7.3.1-**31**
regeneration air	再生空气	air de régénération	7.3.1-**32**
regeneration specific heat input	单位除湿量的热耗量	chaleur de régénération	7.3.1-**33**
regenerator	再生器	régénérateur	7.3.1-**34**
scrubber *air washer*	空气洗涤器	laveur d'air	7.3.1-**35**
silica aerogel	带孔硅胶	aérosilicagel	7.3.1-**36**
spinning disc humidifier	转盘式加湿器	humidificateur à disque tournant	7.3.1-**37**
spray chamber	喷水室	chambre de pulvérisation	7.3.1-**38**
spray nozzle	喷淋喷嘴	pulvérisateur	7.3.1-**39**
spray-type air washer	喷淋式空气洗涤器	laveur d'air à pulvérisation	7.3.1-**40**
surface dehumidifier	表面除湿器	déshumidificateur à action de surface	7.3.1-**41**

SECTION 7.3.2 *Air circulation and distribution: specific equipment*	节 **7.3.2** 空气循环及输配：专用设备	**SOUS-CHAPITRE 7.3.2** *Circulation et distribution de l'air : équipements spécifiques*	
air diffuser	散流器	diffuseur d'air	7.3.2-**1**
air duct *trunking (1)* ○	风道 风管	gaine d'air *conduit d'air* ◐	7.3.2-**2**
air grille *grille*	格栅风口	grille à air	7.3.2-**3**
air-heating fan-coil unit	空气加热风机盘管机组	ventiloconvecteur en mode de chauffage	7.3.2-**4**
air terminal device	空气末端装置	bouche d'air	7.3.2-**5**
blending box *mixing box* *mixing unit*	混合箱	boîte de mélange *caisson de mélange* *chambre de mélange* ◐	7.3.2-**6**
butterfly damper	蝶阀	registre papillon	7.3.2-**7**
ceiling diffuser	顶棚散流器	diffuseur plafonnier	7.3.2-**8**
ceiling outlet	顶棚出风口	bouche de plafond *ouverture de plafond* ◐	7.3.2-**9**
coefficient of discharge	流量系数	coefficient de débit	7.3.2-**10**
core area ◐ *cross area*	格栅风口截面积	section totale	7.3.2-**11**
cross area *core area* ◐	格栅风口截面积	section totale	7.3.2-**12**
damper	风阀	registre	7.3.2-**13**
deflector *turning vane*	导流板	déflecteur	7.3.2-**14**
diffuser	散流器	diffuseur	7.3.2-**15**

	ENGLISH	汉语	FRANÇAIS
7.3.2-**16**	duct	风管 风道	gaine *conduit* ◐
7.3.2-**17**	duct distribution	风道输配	répartition de l'air par gaines
7.3.2-**18**	duct fittings	风管配件	composants de gaine
7.3.2-**19**	ductwork *trunking (2)* ○	风道系统	réseau de gaines *système de gaines* ◐
7.3.2-**20**	equal friction method duct sizing	等摩阻法	dimensionnement des conduits par la méthode d'égal frottement
7.3.2-**21**	equalizing damper	平衡风门	registre d'égalisation
7.3.2-**22**	exhaust opening	排气孔 吸风口	bouche de sortie d'air *orifice de sortie d'air* ◐
7.3.2-**23**	false ceiling *intermediate ceiling*	通风顶棚 夹层平顶	faux plafond
7.3.2-**24**	false floor	通风（夹层）地板	faux plancher
7.3.2-**25**	fire-and-smoke damper	防烟火风阀	clapet coupe-feu
7.3.2-**26**	fixed guard	紧固装置	protecteur fixe
7.3.2-**27**	flexible duct	柔性风管	conduit flexible
7.3.2-**28**	free area	（格栅风口）净面积	section libre de passage d'air
7.3.2-**29**	grille *air grille*	格栅风口	grille à air
7.3.2-**30**	guard	防护罩	protecteur
7.3.2-**31**	in-duct method	风管内取样法	essai en conduit ◐
7.3.2-**32**	induction unit	诱导器	éjecto-convecteur ◐
7.3.2-**33**	intermediate ceiling *false ceiling*	通风顶棚 夹层平顶	faux plafond
7.3.2-**34**	iris damper	光圈风阀	registre à iris
7.3.2-**35**	linear air diffuser	线性空气散流器	diffuseur d'air linéaire
7.3.2-**36**	linear grille	条型风口	grille linéaire
7.3.2-**37**	louvre	百叶窗	persienne *louvre (marine)* ◐
7.3.2-**38**	mixing box *blending box* *mixing unit*	混合箱	boîte de mélange *caisson de mélange* *chambre de mélange* ◐
7.3.2-**39**	mixing unit *blending box* *mixing box*	混合箱	boîte de mélange *caisson de mélange* *chambre de mélange* ◐
7.3.2-**40**	modulating damper	调节型风阀	registre de réglage
7.3.2-**41**	multi-leaf damper	多叶风阀	registre à persiennes
7.3.2-**42**	non-ducted air conditioner	无风管空调器	climatiseur non raccordé
7.3.2-**43**	non-return damper	止回风门	registre à sens unique
7.3.2-**44**	nozzle outlet	送风喷口	buse de diffusion
7.3.2-**45**	outside-air intake duct	室外空气进风道	gaine de renouvellement d'air *gaine d'air extérieur* ◐
7.3.2-**46**	perforated ceiling *ventilated ceiling* ◐	孔板吊顶	plafond perforé
7.3.2-**47**	plenum chamber *plenum space*	静压箱 稳压层	plenum *chambre de répartition d'air* ○

ENGLISH	汉语	FRANÇAIS	
plenum space *plenum chamber*	静压箱 稳压层	plenum *chambre de répartition d'air* ○	7.3.2-**48**
rain louvre *weather louvre* ◐	防雨百页通风窗	grille d'air extérieur contre les intempéries	7.3.2-**49**
register ◐	百叶型风口	grille à registre	7.3.2-**50**
shut-off damper	防火阀	registre d'isolement	7.3.2-**51**
single-leaf damper	单叶风阀	registre à volet	7.3.2-**52**
slide damper	插板风阀	registre à glissières *registre (à) guillotine*	7.3.2-**53**
slot diffuser	条缝型风口	diffuseur linéaire	7.3.2-**54**
splitter	导流板	aubage directeur	7.3.2-**55**
static regain method duct sizing	静压复得法风道计算	dimensionnement des conduits par la méthode de regain de pression statique	7.3.2-**56**
trunking (1) ○ *air duct*	风道 风管	gaine d'air *conduit d'air* ◐	7.3.2-**57**
trunking 2) ○ *ductwork*	风道系统	réseau de gaines *système de gaines* ◐	7.3.2-**58**
turning vane *deflector*	导流板	déflecteur	7.3.2-**59**
velocity reduction method duct sizing	速度递减法风道计算	dimensionnement des conduits par la méthode de réduction des vitesses	7.3.2-**60**
ventilated ceiling ◐ *perforated ceiling*	孔板吊顶	plafond perforé	7.3.2-**61**
weather louvre ◐ *rain louvre*	防雨百页通风窗	grille d'air extérieur contre les intempéries	7.3.2-**62**

SECTION 7.3.3 *Air quality:* *specific equipment*	节 **7.3.3** 空气质量：专用设备	**SOUS-CHAPITRE 7.3.3** *Qualité de l'air:* *équipements spécifiques*	
absolute filter	绝对（高效）过滤器	filtre absolu *filtre ultrafin* ◐ *ultrafiltre* ◐	7.3.3-**1**
air cleaner *air filter*	空气净化器 空气过滤器	filtre d'air *filtre à air*	7.3.3-**2**
air filter *air cleaner*	空气净化器 空气过滤器	filtre d'air *filtre à air*	7.3.3-**3**
automatic roll filter	自动卷绕式过滤器	filtre à déroulement automatique	7.3.3-**4**
brush filter	滤网	filtre à brosses	7.3.3-**5**
carbon filter	活性碳过滤器	filtre à charbon actif	7.3.3-**6**
cartridge filter *cellular filter*	单元式滤芯	filtre à alvéoles *filtre à panneaux* ◐ *filtre à cellules* ◐	7.3.3-**7**
cellular filter *cartridge filter*	单元式滤芯	filtre à alvéoles *filtre à panneaux* ◐ *filtre à cellules* ◐	7.3.3-**8**
disposable air filter	一次性空气过滤器	filtre jetable *filtre à usage unique* ◐	7.3.3-**9**

	ENGLISH	汉语	FRANÇAIS
7.3.3-**10**	dry-layer filter	干式过滤器	filtre sec *filtre à couche sèche* ◐
7.3.3-**11**	dust eliminator	除尘器	dépoussiéreur
7.3.3-**12**	dust extracting plant	除尘装置	installation de filtrage *installation de dépoussiérage*
7.3.3-**13**	dust-spot procedures for testing air-cleaning devices	测试空气洁净设备的点尘法	rendement opacimétrique
7.3.3-**14**	electric precipitator *electrostatic filter* *electrostatic precipitator*	静电除尘器 静电过滤器	électrofiltre *filtre électrostatique* *séparateur électrostatique* ◐
7.3.3-**15**	electrostatic filter *electric precipitator* *electrostatic precipitator*	静电除尘器 静电过滤器	électrofiltre *filtre électrostatique* *séparateur électrostatique* ◐
7.3.3-**16**	electrostatic precipitator *electric precipitator* *electrostatic filter*	静电除尘器 静电过滤器	électrofiltre *filtre électrostatique* *séparateur électrostatique* ◐
7.3.3-**17**	fabric filter	织物过滤器	filtre textile
7.3.3-**18**	fibre-pad filter ◐ *fibrous filter*	纤维过滤器	filtre à matière fibreuse *filtre à masse fibreuse* ◐
7.3.3-**19**	fibrous filter *fibre-pad filter* ◐	纤维过滤器	filtre à matière fibreuse *filtre à masse fibreuse* ◐
7.3.3-**20**	filter cartridge ◐ *filter unit* *filter cell* ◐ *filter element* ◐	过滤单元	cartouche filtrante *élément d'un filtre* ◐
7.3.3-**21**	filter cell ◐ *filter unit* *filter cartridge* ◐ *filter element* ◐	过滤单元	cartouche filtrante *élément d'un filtre* ◐
7.3.3-**22**	filter element ◐ *filter unit* *filter cartridge* ◐ *filter cell* ◐	过滤单元	cartouche filtrante *élément d'un filtre* ◐
7.3.3-**23**	filter unit *filter cartridge* ◐ *filter cell* ◐ *filter element* ◐	过滤单元	cartouche filtrante *élément d'un filtre* ◐
7.3.3-**24**	fine filter	细过滤器	filtre fin
7.3.3-**25**	fume cupboard	通风柜	hotte (de laboratoire) *sorbonne*
7.3.3-**26**	gravimetric yield	过滤器重量效率	rendement gravimétrique
7.3.3-**27**	HEPA filter	高效微粒过滤器	filtre HEPA
7.3.3-**28**	impact filter	惯性过滤器	filtre à chocs *filtre à inertie* *séparateur à chocs* ◐
7.3.3-**29**	ionizator	离子发生器	ionisateur
7.3.3-**30**	laminar flow	层流	flux laminaire
7.3.3-**31**	medium-efficacy air filter	中效空气过滤器	filtre grossier
7.3.3-**32**	moving curtain filter ◐ *roll filter*	卷绕式过滤器	filtre à déroulement
7.3.3-**33**	ozone	臭氧	ozone

ENGLISH	汉语	FRANÇAIS	
ozoniser	臭氧发生器	ozoniseur	7.3.3-**34**
particle meter	粒子计数器	compteur de particules	7.3.3-**35**
primary filter	初效过滤器	préfiltre	7.3.3-**36**
roll filter *moving curtain filter* ◐	卷绕式过滤器	filtre à déroulement	7.3.3-**37**
sorption filter	吸附过滤器	filtre à sorption	7.3.3-**38**
terminal filter	终端过滤器	filtre terminal	7.3.3-**39**
ULPA filter	超细粒子过滤器	filtre ULPA	7.3.3-**40**
viscous filter	粘附式过滤器	filtre à imprégnation visqueuse	7.3.3-**41**

SECTION 7.4 *Ventilation*	节 **7.4** 通风	**SOUS-CHAPITRE 7.4** *Ventilation*	
aerofoil (blade) fan	轴流风机	ventilateur à aubes profilées	7.4-**1**
axial (flow) fan *propeller fan* ◐	轴流风机	ventilateur hélicoïde *ventilateur axial* ◐	7.4-**2**
backward curved impeller	后弯叶轮	roue à aubes tournées vers l'arrière	7.4-**3**
bifurcated fan	长轴风机	groupe ventilateur "bulbe" *moto-ventilateur "bulbe"*	7.4-**4**
blade	叶片	aube *pale*	7.4-**5**
blower	鼓风机	soufflante	7.4-**6**
casing	风机外壳	enveloppe	7.4-**7**
centrifugal fan	离心风机	ventilateur centrifuge	7.4-**8**
circulating fan	风扇	ventilateur brasseur d'air	7.4-**9**
contra-rotating fan	双叶轮反向轴流风机	ventilateur contrarotatif	7.4-**10**
controlled forced-draught ventilation	受控强制通风	ventilation mécanique contrôlée	7.4-**11**
cross-flow fan	贯流式风机	ventilateur tangentiel	7.4-**12**
double-flux controlled forced-draught ventilation	双向受控强制通风	ventilation mécanique contrôlée double flux	7.4-**13**
double inlet fan	双进口风机	ventilateur à deux ouïes *ventilateur double ouïe* ◐	7.4-**14**
downstream fairing	下游整流器	carénage aval	7.4-**15**
downstream guide vanes	下游导向叶片	aubage redresseur	7.4-**16**
draught plant	气流输送装置	système de ventilation	7.4-**17**
ducted fan	管道风机	ventilateur à enveloppe	7.4-**18**
exhauster ◐ *induced draught fan*	排风机	ventilateur d'extraction *ventilateur aspirant* ◐	7.4-**19**
fan	风机 风扇	ventilateur	7.4-**20**
fan curve	风机性能曲线	courbe caractéristique d'un ventilateur	7.4-**21**
fan inlet	风机入口	ouïe d'aspiration du ventilateur	7.4-**22**
fan outlet	风机出口	ouïe de refoulement du ventilateur	7.4-**23**

	ENGLISH	汉语	FRANÇAIS
7.4-24	fan power	风机功率	puissance du ventilateur
7.4-25	forward curved impeller	前弯叶轮	roue à aubes inclinées vers l'avant
7.4-26	gas-tight fan	气密型风机	ventilateur étanche
7.4-27	guide vane	导流叶片	aube directrice
7.4-28	(guide) vane axial fan	带导流片的轴流风机	ventilateur axial à aubage directeur
7.4-29	hot-gas fan	高温风机	ventilateur pour gaz chauds
7.4-30	ignition-protected fan *spark-resistant fan* ◐	防爆风机	ventilateur antiétincelles
7.4-31	impeller	叶轮	roue
7.4-32	impeller backplate *impeller hub disc* *impeller hub plate*	叶轮毂板	disque arrière (de roue)
7.4-33	impeller hub disc *impeller backplate* *impeller hub plate*	叶轮毂板	disque arrière (de roue)
7.4-34	impeller hub plate *impeller backplate* *impeller hub disc*	叶轮毂板	disque arrière (de roue)
7.4-35	impeller rim ◐ *inlet ring* *impeller shroud* ◐ *wheel cone* ◐	入口锥形环	disque avant (de roue) *collerette (de roue)* ◐
7.4-36	impeller shroud ◐ *inlet ring* *impeller rim* ◐ *wheel cone* ◐	入口锥形环	disque avant (de roue) *collerette (de roue)* ◐
7.4-37	impeller tip diameter	叶轮直径	diamètre de la roue
7.4-38	induced draught fan *exhauster* ◐	排风机	ventilateur d'extraction *ventilateur aspirant* ◐
7.4-39	industrial fan	工业风机	ventilateur industriel
7.4-40	inlet box	入口箱	caisson d'aspiration *coude d'aspiration* ◐
7.4-41	inlet ring *impeller rim* ◐ *impeller shroud* ◐ *wheel cone* ◐	入口锥形环	disque avant (de roue) *collerette (de roue)* ◐
7.4-42	jet fan	射流风机	ventilateur accélérateur *ventilateur relais* ◐
7.4-43	mixed-flow fan	混流风机	ventilateur hélico-centrifuge
7.4-44	multistage fan	多级风机	ventilateur multiétages
7.4-45	paddle-bladed impeller *radial-bladed impeller*	径向叶轮	roue à aubes radiales
7.4-46	partition fan	嵌墙风机	ventilateur de paroi
7.4-47	plate-mounted axial-flow fan	薄型轴流风机	ventilateur hélicoïde monté sur plaque
7.4-48	powered roof ventilator	屋顶风机	tourelle d'extraction *tourelle de ventilation* *ventilateur de toiture* ◐
7.4-49	propeller fan ◐ *axial (flow) fan*	轴流风机	ventilateur hélicoïde *ventilateur axial* ◐
7.4-50	radial-bladed impeller *paddle-bladed impeller*	径向叶轮	roue à aubes radiales

ENGLISH	汉语	FRANÇAIS	
reversible axial-flow fan	可逆轴流风机	ventilateur hélicoïde réversible	7.4-**51**
ring-shaped fan	环形风机	ventilateur annulaire	7.4-**52**
single-flux controlled forced-draught ventilation	单向受控强制通风	ventilation mécanique contrôlée simple flux	7.4-**53**
smoke-ventilating fan	排烟风机	ventilateur de désenfumage	7.4-**54**
spark-resistant fan ◐ *ignition-protected fan*	防爆风机	ventilateur antiétincelles	7.4-**55**
tip clearance	叶端间隙	jeu radial	7.4-**56**
tube axial fan ◐	管道轴流风机	ventilateur (axial) à enveloppe	7.4-**57**
tubular centrifugal fan	管道离心风机	ventilateur centrifugo-axial	7.4-**58**
upstream fairing	上游整流器	carénage amont	7.4-**59**
upstream guide vanes	上游导流片	distributeur	7.4-**60**
vane axial fan	带导流片的轴流风机	ventilateur hélicoïde à aubes directrices	7.4-**61**
velocity triangle	速度三角形	triangle des vitesses	7.4-**62**
ventilator	通风器	aérateur	7.4-**63**
wet-gas fan	湿式气体风机	ventilateur pour gas humides	7.4-**64**
wheel cone ◐ *inlet ring* *impeller rim* ◐ *impeller shroud* ◐	入口锥形环	disque avant (de roue) *collerette (de roue)* ◐	7.4-**65**

SECTION 7.5 *Packaged and split air-conditioning units*	节 **7.5** 整体式及分体式空气调节机组	**SOUS-CHAPITRE 7.5** *Installations de conditionnement d'air monoblocs ou à éléments séparés*	
air conditioner ◐ *air-conditioning unit*	空气调节器	appareil de conditionnement d'air *climatiseur* *conditionneur d'air* ◐	7.5-**1**
air-conditioning unit *air conditioner* ◐	空气调节器	appareil de conditionnement d'air *climatiseur* *conditionneur d'air* ◐	7.5-**2**
air-cooled air conditioner	风冷式空气调节器	conditionneur d'air à condenseur à air	7.5-**3**
air-handling unit	空气处理机组	caisson de traitement d'air	7.5-**4**
air-terminal unit	空气末端机组	dispositif terminal	7.5-**5**
all-air system	全空气系统	système "tout air"	7.5-**6**
cassette unit	嵌入式机组 盒式机组	unité de type cassette	7.5-**7**
central air-conditioning plant	集中式（中央）空气调节机机组	centrale de conditionnement d'air	7.5-**8**
central fan air-conditioning system	集中式（中央）空气调节系统	système centralisé de conditionnement d'air	7.5-**9**
chilled beam *cold beam*	冷梁	poutre froide *poutre rafraîchissante*	7.5-**10**
cold beam *chilled beam*	冷梁	poutre froide *poutre rafraîchissante*	7.5-**11**
console air conditioner ◐	托架式空气调节器	conditionneur d'air mural	7.5-**12**

	ENGLISH	汉语	FRANÇAIS
7.5-13	district cooling	区域供冷	refroidissement urbain *distribution urbaine de froid* ◐
7.5-14	district heating	区域供热	chauffage urbain
7.5-15	dual-duct air-conditioning system	双风管空气调节系统	système de conditionnement d'air à double conduit
7.5-16	fan-coil unit *fan-convector unit*	风机盘管机组	ventiloconvecteur *batterie ventilée* ○
7.5-17	fan-convector unit *fan-coil unit*	风机盘管机组	ventiloconvecteur *batterie ventilée* ○
7.5-18	four-pipe air-conditioning system	四管制空气调节系统	système de conditionnement d'air à quatre tuyaux
7.5-19	free-blow air conditioner *free delivery-type (air-conditioning) unit*	无风管式空气调节器 无风管式（空气调节） 机组	conditionneur d'air à soufflage direct
7.5-20	free cooling	自由冷却	refroidissement naturel
7.5-21	free delivery-type (air-conditioning) unit *free-blow air conditioner*	无风管式空气调节器 无风管式（空气调节） 机组	conditionneur d'air à soufflage direct
7.5-22	heat-of-light system ○	照明供暖系统	système à éclairage chauffant *système à éclairage intégré*
7.5-23	high-pressure air-conditioning plant	高压空气调节机组	installation de conditionnement d'air à haute pression *installation de conditionnement d'air à grande vitesse*
7.5-24	indoor unit	室内机组	unité intérieure
7.5-25	low-pressure air-conditioning plant	低压空气调节机组	installation de conditionnement d'air à basse pression
7.5-26	modular (air-conditioning) system	模块式（空气调节）系统	installation modulaire (de conditionnement d'air)
7.5-27	multisplit air-conditioning system	多室内机组分体式空调系统	multisplit
7.5-28	night ventilation	夜间通风	ventilation nocturne
7.5-29	outdoor unit	室外机组	unité extérieure
7.5-30	packaged air conditioner *self-contained air-conditioning unit* ◐	整体式空气调节器	conditionneur d'air de type armoire *conditionneur d'air monobloc* ◐
7.5-31	regenerative cooling	再生冷却	refroidissement par récupération
7.5-32	regenerative heating	再生加热	chauffage par récupération
7.5-33	rock-bed regenerative cooling	岩石床再生冷却	refroidissement à récupération sur couches de pierres
7.5-34	roof-top (air-conditioning) unit *roof-top conditioner*	屋顶式（空气调节）机组 屋顶式空气调节器	conditionneur d'air en toiture
7.5-35	roof-top conditioner *roof-top (air-conditioning) unit*	屋顶式（空气调节）机组 屋顶式空气调节器	conditionneur d'air en toiture
7.5-36	room air conditioner (RAC)	房间空调器 (RAC)	conditionneur d'air de pièce *conditionneur d'air unitaire*
7.5-37	self-contained air-conditioning unit ◐ *packaged air conditioner*	整体式空气调节器	conditionneur d'air de type armoire *conditionneur d'air monobloc* ◐
7.5-38	single-duct air-conditioning system	单风管空气调节系统	système de conditionnement d'air à un conduit
7.5-39	split (air-conditioning) system	分体式（空气调节）系统	système split *système (de conditionnement d'air) bibloc* ◐ *conditionneur d'air à condenseur séparé* ○ *conditionneur d'air à deux blocs* ○

ENGLISH	汉语	FRANÇAIS	
split unit	分体空调机组	split	7.5-**40**
three-pipe air-conditioning system	三管制空气调节系统	système de conditionnement d'air à trois tuyaux	7.5-**41**
through-the-wall conditioner ◐	穿墙型空气调节器	conditionneur d'air "à travers le mur"	7.5-**42**
total energy concept ◐ *total energy system*	全能量系统	système à énergie totale	7.5-**43**
total energy system *total energy concept* ◐	全能量系统	système à énergie totale	7.5-**44**
VAV (Variable-Air-Volume) system	变风量 （VAV） 系统	système à débit d'air variable (VAV)	7.5-**45**
VRV (Variable-Refrigerant-Volume) system	变制冷剂流量 （VRV） 系统	système à débit de frigorigène variable) (VRV)	7.5-**46**
water-cooled air conditioner	水冷式空气调节器	conditionneur d'air à condenseur à eau	7.5-**47**
window-air conditioner	窗式空气调节器	conditionneur d'air "type fenêtre"	7.5-**48**
zone air conditioner	区域空气调节器	climatiseur de zone	7.5-**49**

SECTION 7.6 *Air-conditioned spaces*	节 **7.6** 空调房间	**SOUS-CHAPITRE 7.6** *Espaces conditionnés*	
climatic chamber *environmental chamber*	环境试验室	chambre climatique	7.6-**1**
double glazing *dual glazing* ◐	双层玻璃	double-vitrage	7.6-**2**
double window	双层窗	double-fenêtre	7.6-**3**
dual glazing ◐ *double glazing*	双层玻璃	double-vitrage	7.6-**4**
enclosed space *enclosure* ◐	封闭空间	enceinte	7.6-**5**
enclosure ◐ *enclosed space*	封闭空间	enceinte	7.6-**6**
environmental chamber *climatic chamber*	环境试验室	chambre climatique	7.6-**7**
fenestration	透光面积	surface vitrée	7.6-**8**
multiple glazing	多层玻璃	vitrage multiple	7.6-**9**
phytotron	植物气候室	phytotron	7.6-**10**
shading coefficient	遮阳系数	facteur d'écran *facteur d'ombrage*	7.6-**11**
shading device	遮阳装置	pare-soleil *brise-soleil* ◐	7.6-**12**
spot cooling	局部冷却	refroidissement localisé	7.6-**13**

SECTION 7.7 *Clean rooms*	节 **7.7** 洁净室	**SOUS-CHAPITRE 7.7** *Salles blanches*	
alert level	报警浓度	niveau d'alerte	7.7-**1**
biocontamination	生物污染	biocontamination	7.7-**2**

	ENGLISH	汉语	FRANÇAIS
7.7-3	classification	微粒分级	classification
7.7-4	clean room	洁净室	salle blanche *enceinte à empoussiérage contrôlé* *salle propre*
7.7-5	clean space *clean zone*	洁净空间 r洁净区	espace propre *espace à empoussiérage contrôlé*
7.7-6	clean work station	洁净工作台	poste de travail propre
7.7-7	clean zone *clean space*	洁净空间 洁净区	espace propre *espace à empoussiérage contrôlé*
7.7-8	controlled environment	受控环境	environnement maîtrisé
7.7-9	dust-controlled clean rooms: class	控尘洁净室:级别	classe des locaux à empoussièrement contrôlé
7.7-10	fibre	纤维尘	fibre
7.7-11	internal generation of particles	内部粒子生成	génération interne de particules
7.7-12	macroparticle	大微粒	macroparticule
7.7-13	particle	微粒	particule
7.7-14	particle concentration	微粒浓度	concentration de particules
7.7-15	particle size	粒径	taille de particule
7.7-16	particle size distribution	粒径分布	distribution granulométrique
7.7-17	ultrafine particle	超细微粒	particule ultrafine
7.7-18	viable particle	生物微粒	particule viable
7.7-19	zone at risk	污染风险区	zone à risque

章 8. 热泵

◐ 容许（或默许）术语

○ 过时术语

ENGLISH	汉语	FRANÇAIS	
CHAPTER 8 *Heat pumps*	章 **8** 热泵	**CHAPITRE 8** *Pompes à chaleur*	
absorption heat pump	吸收式热泵	pompe à chaleur à absorption	**8-1**
adsorption heat pump	吸附式热泵	pompe à chaleur à adsorption	**8-2**
air-source heat pump	空气源热泵	pompe à chaleur sur l'air	**8-3**
air-to-air heat pump	空气-空气热泵	pompe à chaleur air-air	**8-4**
air-to-water heat pump	空气-水热泵	pompe à chaleur air-eau	**8-5**
brine-to-air heat pump ◐ *water-to-air heat pump*	水-空气热泵	pompe à chaleur eau-air	**8-6**
brine-to-water heat pump	盐水-水热泵	pompe à chaleur saumure-eau	**8-7**
chemical heat pump	化学热泵	pompe à chaleur à réaction chimique	**8-8**
closed-loop ground-source heat pump ◐ *ground-source heat pump* *ground-coupled heat pump* *geothermal heat pump* ◐ *ground-loop heat pump* ◐	大地耦合（地源）热泵	pompe à chaleur couplée au sol	**8-9**
coefficient of performance (of a heat pump)	(热泵的)性能系数	coefficient de performance (d'une pompe à chaleur) *coefficient d'efficacité calorifique (d'une pompe à chaleur)* ◐ *coefficient d'amplification (d'une pompe à chaleur)* ◐	**8-10**
compression heat pump	蒸气压缩式热泵	pompe à chaleur à compression	**8-11**
direct-expansion ground-coupled heat pump	直接膨胀式地源热泵	pompe à chaleur sol-air ou sol-eau à évaporation directe	**8-12**
geothermal heat pump ◐ *ground-source heat pump* *ground-coupled heat pump* *closed-loop ground-source heat pump* ◐ *ground-loop heat pump* ◐	大地耦合（地源）热泵	pompe à chaleur couplée au sol	**8-13**
ground-coupled heat pump *ground-source heat pump* *closed loop ground-source heat pump* ◐ *geothermal heat pump* ◐ *ground-loop heat pump* ◐	大地耦合（地源）热泵	pompe à chaleur couplée au sol	**8-14**
ground-loop heat pump ◐ *ground-source heat pump* *ground-coupled heat pump* *closed-loop ground-source heat pump* ◐ *geothermal heat pump* ◐	大地耦合（地源）热泵	pompe à chaleur couplée au sol	**8-15**
ground-source heat pump *ground-coupled heat pump* *closed-loop ground-source heat pump* ◐ *geothermal heat pump* ◐ *ground-loop heat pump* ◐	大地耦合（地源）热泵	pompe à chaleur couplée au sol	**8-16**
ground-water heat pump ◐	地下水源热泵	pompe à chaleur sur eau souterraine	**8-17**
heat output	供热量	puissance thermique	**8-18**
heat pump	热泵	pompe à chaleur	**8-19**
heat pump boiler ◐ *heat pump water heater*	热泵型热水器	pompe à chaleur pour chauffage d'eau	**8-20**
heat pump water heater *heat pump boiler* ◐	热泵型热水器	pompe à chaleur pour chauffage d'eau	**8-21**

	ENGLISH	汉语	FRANÇAIS
8-22	heat-recovery heat pump	热回收热泵	pompe à chaleur pour récupération de chaleur
8-23	heat transformer *temperature amplifier* ◐	吸收式升温器	transformateur de chaleur
8-24	heating energy	供热量	énergie thermique
8-25	heating seasonal performance factor *HSPF*	采暖季节性能系数	coefficient de performance moyen saisonnier
8-26	HSPF *heating seasonal performance factor*	采暖季节性能系数	coefficient de performance moyen saisonnier
8-27	mechanical vapour recompression	机械驱动蒸气再压缩	recompression mécanique de vapeur
8-28	primary energy ratio	一次能源能效比	efficacité rapportée à l'énergie primaire
8-29	reverse-cycle heating ◐ *thermodynamic heating* ○	逆循环供热	chauffage par cycle inversé ◐ *chauffage par cycle frigorifique* ◐ *chauffage thermodynamique* ◐
8-30	solar-assisted heat pump ◐	太阳能辅助热泵	pompe à chaleur assistée par l'énergie solaire
8-31	surface-water heat pump ◐	地表水源热泵	pompe à chaleur sur eau de surface
8-32	temperature amplifier ◐ *heat transformer*	吸收式升温器	transformateur de chaleur
8-33	thermal vapour recompression	热驱动蒸汽再压缩	thermocompression de vapeur
8-34	thermodynamic heating ○ *reverse cycle heating* ◐	逆循环供热	chauffage par cycle inversé ◐ *chauffage par cycle frigorifique* ◐ *chauffage thermodynamique* ◐
8-35	thermoelectric heat pump	热电式热泵	pompe à chaleur thermoélectrique
8-36	ventilation air heat pump ◐	通风排气源热泵	pompe à chaleur sur air extrait
8-37	water-loop heat pump	水环热泵	pompe à chaleur sur boucle d'eau
8-38	water-source heat pump	水源热泵	pompe à chaleur sur eau
8-39	water-to-air heat pump *brine-to-air heat pump* ◐	水-空气热泵	pompe à chaleur eau-air
8-40	water-to-water heat pump	水-水热泵	pompe à chaleur eau-eau

章 9. 低温学

◐ 容许（或默许）术语

○ 过时术语

ENGLISH	汉语	FRANÇAIS	
SECTION 9.1 *Cryophysics* **SECTION 9.1.1** *Cryogenics and cryoengineering*	节 **9.1** 低温物理学 节 **9.1.1** 低温技术与低温工程	**SOUS-CHAPITRE 9.1** *Cryophysique* **SOUS-CHAPITRE 9.1.1** *Cryogénie et cryotechnique*	
adiabatic demagnetization	绝热去磁	désaimantation adiabatique	9.1.1-**1**
air fractionation ○ *air separation*	空气分离	séparation de l'air	9.1.1-**2**
air separation *air fractionation* ○	空气分离	séparation de l'air	9.1.1-**3**
bubble chamber	气泡室	chambre à bulles	9.1.1-**4**
cold box	冷箱	boîte froide	9.1.1-**5**
cryoalternator	低温交流发电机 低温同步发电机	cryoalternateur	9.1.1-**6**
cryocable	低温电缆	cryocâble	9.1.1-**7**
cryochemistry	低温化学	cryochimie	9.1.1-**8**
cryoconductor	低温导线	cryoconducteur (subst.) *hyperconducteur*	9.1.1-**9**
cryocooling *cryogenic cooling*	低温冷却	cryorefroidissement	9.1.1-**10**
cryoelectric *cryoelectrical*	低温电（的）	cryoélectrique	9.1.1-**11**
cryoelectrical *cryoelectric*	低温电（的）	cryoélectrique	9.1.1-**12**
cryoelectronics	低温电子学	cryoélectronique	9.1.1-**13**
cryoelectrotechnics	低温电工学	cryoélectrotechnique	9.1.1-**14**
cryoengineering *cryogenic engineering* ◐	低温工程学	cryotechnique *ingénierie cryogénique* *technique cryogénique*	9.1.1-**15**
cryogen *cryogenic fluid*	低温流体 低温制冷剂	cryogène *fluide cryogénique*	9.1.1-**16**
cryogenic	低温（的）	cryogénique	9.1.1-**17**
cryogenic bath	低温液池	bain cryogénique	9.1.1-**18**
cryogenic cooling *cryocooling*	低温冷却	cryorefroidissement	9.1.1-**19**
cryogenic engineering ◐ *cryoengineering*	低温工程学	cryotechnique *ingénierie cryogénique* *technique cryogénique*	9.1.1-**20**
cryogenic equipment	低温器材	matériel cryogénique *appareil cryogénique* *équipement cryogénique* ○	9.1.1-**21**
cryogenic fluid *cryogen*	低温流体 低温制冷剂	cryogène *fluide cryogénique*	9.1.1-**22**
cryogenic liquid	低温液体	liquide cryogénique	9.1.1-**23**
cryogenic plant	低温制冷设备	installation cryogénique	9.1.1-**24**
cryogenic process	低温制冷过程 低温制冷工艺	procédé cryogénique	9.1.1-**25**
cryogenic propellant ◐ *cryopropellant*	低温推进剂	ergol cryogénique	9.1.1-**26**

	ENGLISH	汉语	FRANÇAIS
9.1.1-**27**	cryogenic refrigerator ◐ *cryorefrigerator*	低温制冷机	cryoréfrigérateur
9.1.1-**28**	cryogenic storage	低温储存	stockage cryogénique
9.1.1-**29**	cryogenic (storage) vessel	低温储液容器	réservoir cryogénique
9.1.1-**30**	cryogenic tanker	低温（液体）槽船	citerne cryogénique
9.1.1-**31**	cryogenic technique ◐ *cryotechnique*	低温实验技术	cryotechnique (expérimentale)
9.1.1-**32**	cryogenic technology ◐ *cryotechnology*	低温工艺	cryotechnique (procédés)
9.1.1-**33**	cryogenic valve ◐ *cryovalve*	低温阀	cryovanne *vanne cryogénique*
9.1.1-**34**	cryogenics *cryology*	低温学	cryogénie
9.1.1-**35**	cryoliquefier	低温液化器	cryoliquéfacteur
9.1.1-**36**	cryology *cryogenics*	低温学	cryogénie
9.1.1-**37**	cryomachining	低温机加工	cryo-usinage
9.1.1-**38**	cryomagnet	低温磁体	cryoaimant
9.1.1-**39**	cryomicroscope	低温显微镜	cryomicroscope
9.1.1-**40**	cryomotor	低温电动机	cryomoteur
9.1.1-**41**	cryophysics	低温物理学	cryophysique
9.1.1-**42**	cryopropellant *cryogenic propellant* ◐	低温推进剂	ergol cryogénique
9.1.1-**43**	cryopump	低温真空泵	cryopompe
9.1.1-**44**	cryorefrigerator *cryogenic refrigerator* ◐	低温制冷机	cryoréfrigérateur
9.1.1-**45**	cryoresistive	低温低阻导电（的）	cryoconducteur (adj.)
9.1.1-**46**	cryostat	低温恒温器	cryostat
9.1.1-**47**	cryotechnique *cryogenic technique* ◐	低温实验技术	cryotechnique (expérimentale)
9.1.1-**48**	cryotechnology *cryogenic technology* ◐	低温工艺	cryotechnique (procédés)
9.1.1-**49**	cryotemperature	低温（温度）	cryotempérature *température cryogénique*
9.1.1-**50**	cryotransformer	低温变压器	cryotransformateur
9.1.1-**51**	cryotrap	低温冷阱	cryopiège
9.1.1-**52**	cryotron ○	低温电子管	cryotron
9.1.1-**53**	cryovalve *cryogenic valve* ◐	低温阀	cryovanne *vanne cryogénique*
9.1.1-**54**	Dewar (vessel) *vacuum flask* ○	杜瓦瓶	(vase) Dewar
9.1.1-**55**	dilution refrigerator	稀释制冷机	réfrigérateur à dilution
9.1.1-**56**	expansion method *Simon's expansion method*	西蒙膨胀法	cryorefroidissement par détente *détente de Simon*
9.1.1-**57**	gas separation unit	低温气体分离装置	appareil de séparation des gaz
9.1.1-**58**	helium desorption method	退吸附法	cryorefroidissement par désorption

ENGLISH	汉语	FRANÇAIS	
magnetic cooling ◐	磁制冷	refroidissement magnétique	9.1.1-**59**
nuclear alignment	核排列	alignement nucléaire	9.1.1-**60**
nuclear cooling	核磁制冷	refroidissement nucléaire	9.1.1-**61**
nuclear orientation	核取向	orientation nucléaire	9.1.1-**62**
nuclear polarization	核极化	polarisation nucléaire	9.1.1-**63**
phonon drag	声子曳引	entraînement des phonons	9.1.1-**64**
Simon's expansion method *expansion method*	西蒙膨胀法	cryorefroidissement par détente *détente de Simon*	9.1.1-**65**
supercritical	超临界（的）	supercritique	9.1.1-**66**
vacuum flask ○ *Dewar (vessel)*	杜瓦瓶	(vase) Dewar	9.1.1-**67**

SECTION 9.1.2 *Liquid helium*	节 **9.1.2** 液氦	**SOUS-CHAPITRE 9.1.2** *Hélium liquide*	
creep rate (He II)	He II 膜爬移速率	vitesse d'écoulement en film *vitesse de grimpage* ◐	9.1.2-**1**
critical velocity (He II)	He II 临界流速	vitesse critique (He II)	9.1.2-**2**
fountain effect (He II) *thermomechanical effect* ◐	喷泉效应	effet fontaine *effet thermomécanique* ◐	9.1.2-**3**
fourth sound	第四声	quatrième son	9.1.2-**4**
helium film (He II) *Rollin film* ◐	He II 膜 Rollin 膜	film d'hélium *film de Rollin* ◐	9.1.2-**5**
lambda leak (He II) ◐	λ 漏（孔）	fuite lambda	9.1.2-**6**
lambda line (He-4)	λ 线	courbe lambda (He-4)	9.1.2-**7**
lambda point (He-4)	λ 点	point lambda (He-4)	9.1.2-**8**
parafluidity ○ *parasuperfluidity*	予超流动性	parasuperfluidité *parafluidité* ○	9.1.2-**9**
parasuperfluidity *parafluidity* ○	予超流动性	parasuperfluidité *parafluidité* ○	9.1.2-**10**
Rollin film ◐ *helium film (He II)*	He II 膜 Rollin 膜	film d'hélium *film de Rollin* ◐	9.1.2-**11**
rotons	旋子	rotons	9.1.2-**12**
second sound (He II)	第二声（He II）	deuxième son	9.1.2-**13**
superfluid flow (He II)	He II 超流流动	écoulement superfluide	9.1.2-**14**
superfluidity (He II)	He II 超流动性	superfluidité	9.1.2-**15**
superleak (He II)	超流通道 超漏	superfuite	9.1.2-**16**
thermomechanical effect ◐ *fountain effect (He II)*	喷泉效应	effet fontaine *effet thermomécanique* ◐	9.1.2-**17**
third sound	第三声	troisième son	9.1.2-**18**
two-fluid model (He II)	二流体模型	modèle à deux fluides	9.1.2-**19**
zeroth sound	第零声	son zéro	9.1.2-**20**

	ENGLISH	汉语	FRANÇAIS
	SECTION 9.1.3 *Superconductivity*	节 **9.1.3** 超导	**SOUS-CHAPITRE 9.1.3** *Supraconductivité*
9.1.3-**1**	AC Josephson effect	交流约瑟夫森效应	effet Josephson alternatif
9.1.3-**2**	adiabatic stabilization	绝热稳定化	stabilisation adiabatique
9.1.3-**3**	coherence length	相干长度	longueur de cohérence
9.1.3-**4**	critical current	临界电流	courant critique
9.1.3-**5**	critical field (H_c) *thermodynamical critical field* ◐	临界磁场强度	champ critique (H_c) *champ critique thermodynamique* ◐
9.1.3-**6**	critical temperature (superconductor)	临界温度	température critique (supraconducteur)
9.1.3-**7**	cryostabilization	低温稳定化	cryostabilisation *stabilisation cryogénique* ◐
9.1.3-**8**	DC Josephson effect	直流约瑟夫森效应	effet Josephson continu
9.1.3-**9**	degradation (superconductor)	退化	dégradation (supraconducteur)
9.1.3-**10**	dynamic stabilization	动力学稳定化	stabilisation dynamique
9.1.3-**11**	filamentary superconductor	多丝超导线	supraconducteur filamentaire
9.1.3-**12**	flux flow	磁通流动	écoulement de flux
9.1.3-**13**	flux jump	磁通跳跃	saut de flux
9.1.3-**14**	fluxoid	类磁通	fluxoïde
9.1.3-**15**	fluxon	磁通涡旋线	fluxon
9.1.3-**16**	frozen-in flux ◐ *trapped flux*	俘获磁通 冻结磁通	flux piégé
9.1.3-**17**	intermediate state	中间态	état intermédiaire
9.1.3-**18**	intrinsic stabilization	本征稳定化	stabilisation intrinsèque
9.1.3-**19**	Josephson effects	约瑟夫森效应	effets Josephson
9.1.3-**20**	lower critical field (H_{c1})	下临界磁场强度	champ critique inférieur (H_{c1})
9.1.3-**21**	magnetic penetration depth *penetration depth* ◐	穿透深度	profondeur de pénétration
9.1.3-**22**	maximum recovery current *minimum propagating current*	最小传播电流	courant minimal de propagation résistive *courant maximal de récupération*
9.1.3-**23**	Meissner state	迈斯纳态	état de Meissner
9.1.3-**24**	minimum propagating current *maximum recovery current*	最小传播电流	courant minimal de propagation résistive *courant maximal de récupération*
9.1.3-**25**	minimum propagating zone *MPZ*	最小传播区	zone résistive minimale de propagation
9.1.3-**26**	mixed state	混合态	état mixte
9.1.3-**27**	MPZ *minimum propagating zone*	最小传播区	zone résistive minimale de propagation
9.1.3-**28**	normal state	正常态	état normal
9.1.3-**29**	penetration depth ◐ *magnetic penetration depth*	穿透深度	profondeur de pénétration
9.1.3-**30**	persistent current	持续电流	courant persistant
9.1.3-**31**	proximity effect	邻近效应	effet de proximité
9.1.3-**32**	SQUID (superconducting quantum interference device)	超导量子干涉器件	SQUID *interféromètre quantique supraconducteur*

ENGLISH	汉语	FRANÇAIS	
stabilization	稳定化	stabilisation	9.1.3-**33**
superconducting state *superconductive state* ◐	超导态	état supraconducteur	9.1.3-**34**
superconduction ◐ *superconductivity*	超导电性	supraconductivité *supraconduction*	9.1.3-**35**
superconductive state ◐ *superconducting state*	超导态	état supraconducteur	9.1.3-**36**
superconductivity *superconduction* ◐	超导电性	supraconductivité *supraconduction*	9.1.3-**37**
superconductor	超导体	supraconducteur	9.1.3-**38**
thermodynamical critical field ◐ *critical field (H_c)*	临界磁场强度	champ critique (H_c) *champ critique thermodynamique* ◐	9.1.3-**39**
transition temperature	转变温度	température de transition	9.1.3-**40**
trapped flux *frozen-in flux* ◐	俘获磁通 冻结磁通	flux piégé	9.1.3-**41**
type I superconductor	第一类超导体	supraconducteur de type I	9.1.3-**42**
type II superconductor	第二类超导体	supraconducteur de type II	9.1.3-**43**
upper critical field (H_{c2})	上临界磁场强度	champ critique supérieur (H_{c2})	9.1.3-**44**

SECTION 9.1.4 *Liquefied gases*	节 **9.1.4** 液化气	**SOUS-CHAPITRE 9.1.4** *Gaz liquéfiés*	
air-separation plant	空气分离装置	installation de séparation d'air	9.1.4-**1**
ambient air vaporizer	环境空气气化器	vaporiseur dans l'air ambiant *vaporisateur dans l'air ambiant*	9.1.4-**2**
argon column *argon side column* ◐ *side-arm column* ◐	氩塔	colonne d'argon	9.1.4-**3**
argon side column ◐ *argon column* *side-arm column* ◐	氩塔	colonne d'argon	9.1.4-**4**
balance stream ◐ *trumpler pass*	平衡流	flux d'équilibrage	9.1.4-**5**
ballasting	稳定发热量	ballastage	9.1.4-**6**
base load (LNG plant)	LNG 基本负荷装置	capacité de production nominale (installation de GNL)	9.1.4-**7**
bayonet joint	承插式接头	raccord à baïonnette	9.1.4-**8**
boil-off	逸气	pertes par évaporation	9.1.4-**9**
boil-off rate (BOR) *BOR* *evaporation rate* *NER* *net evaporation rate (NER)*	蒸发率 (BOR) 蒸发率	taux d'évaporation *taux de pertes par évaporation*	9.1.4-**10**
BOR *boil-off rate (BOR)* *evaporation rate* *NER* *net evaporation rate (NER)*	蒸发率 (BOR) 蒸发率	taux d'évaporation *taux de pertes par évaporation*	9.1.4-**11**
brazed-aluminium heat exchanger *plate-and-fin heat exchanger* ◐	钎焊铝换热器 板翅式换热器	échangeur de chaleur en aluminium brasé	9.1.4-**12**

	ENGLISH	汉语	FRANÇAIS
9.1.4-**13**	bund wall	防护墙	cuvette de rétention *mur de rétention* ◐
9.1.4-**14**	cascade cycle	复叠式循环	cycle à cascade *procédé à cascade* ◐
9.1.4-**15**	Claude cycle ◐ *medium-pressure cycle*	中压循环 克劳德循环中压装置	cycle moyenne pression
9.1.4-**16**	CLOX *crude liquid oxygen (CLOX)*	原氧液	oxygène liquide brut
9.1.4-**17**	cold recovery	冷量回收	récupération de froid
9.1.4-**18**	compander	增压透平膨胀机	unité de compression/détente
9.1.4-**19**	condenser-reboiler	冷凝蒸发器	condenseur-vaporiseur
9.1.4-**20**	crude liquid oxygen (CLOX) *CLOX*	原氧液	oxygène liquide brut
9.1.4-**21**	cryogenic storage tank vacuum-insulated tank	真空绝热罐 低温贮罐	réservoir isolé sous vide
9.1.4-**22**	customer station *vaporization station* *satellite station*	气化装置	poste de vaporisation
9.1.4-**23**	denitrogenation	除氮	dénitrogénation *désazotation*
9.1.4-**24**	double-column plant	双塔装置	unité à double colonne
9.1.4-**25**	downflow condenser-reboiler	下降流冷凝蒸发器	condenseur-vaporiseur à flux descendant
9.1.4-**26**	EOR (enhanced oil recovery)	提高原油采收率法	récupération assistée
9.1.4-**27**	evaporation rate *boil-off rate (BOR)* *BOR* *NER* *net evaporation rate (NER)*	蒸发率 （BOR） 蒸发率	taux d'évaporation *taux de pertes par évaporation*
9.1.4-**28**	expander ◐ *expansion turbine*	透平膨胀机	turbine à expansion *turbine de détente*
9.1.4-**29**	expansion bellows	膨胀波纹管	soufflet d'expansion
9.1.4-**30**	expansion engine	往复膨胀机	moteur à expansion
9.1.4-**31**	expansion turbine *expander* ◐	透平膨胀机	turbine à expansion *turbine de détente*
9.1.4-**32**	flat-bottom tank	平底罐	réservoir à fond plat
9.1.4-**33**	GAN *gaseous nitrogen (GAN)*	氮气 （GAN） 氮气	azote gazeux
9.1.4-**34**	gaseous nitrogen (GAN) *GAN*	氮气 （GAN） 氮气	azote gazeux
9.1.4-**35**	gaseous oxygen (GOX) *GOX*	氧气 （GOX） 氧气	oxygène gazeux
9.1.4-**36**	GOX *gaseous oxygen (GOX)*	氧气 （GOX） 氧气	oxygène gazeux
9.1.4-**37**	heat leakage	热渗漏	fuite thermique *déperdition de chaleur* ◐
9.1.4-**38**	Heylandt cycle *high-pressure cycle*	高压循环 海兰德循环	cycle de Heylandt
9.1.4-**39**	high-pressure column *HP column* ◐ *lower column* ◐	高压塔 下塔	colonne haute pression

ENGLISH	汉语	FRANÇAIS	
high-pressure cycle *Heylandt cycle*	高压循环 海兰德循环	cycle de Heylandt	9.1.4-**40**
HP column ◐ *high-pressure column* *lower column* ◐	高压塔 下塔	colonne haute pression	9.1.4-**41**
inner cascade cycle *mixed-refrigerant cycle* *mixed-refrigerant liquefier*	内复叠式循环 内复叠式循环液化器	cycle à fluide frigorigène mixte	9.1.4-**42**
internal-compression plant ◐ *liquid-pump plant*	液氧泵式装置 内压缩式装置	unité de séparation d'air à pompe *unité de séparation d'air à compression interne* ◐	9.1.4-**43**
Kellog cycle ◐ *low-pressure cycle* *low-pressure liquefier*	低压循环	cycle basse pression	9.1.4-**44**
Le Rouget cycle	拉鲁日循环	cycle de Le Rouget	9.1.4-**45**
LIN *liquid nitrogen (LIN)*	液氮 (LIN) 液氮	azote liquide	9.1.4-**46**
Linde column ◐ *single-column plant*	单塔装置	unité à une seule colonne	9.1.4-**47**
Linde cycle	林德循环	cycle de Linde	9.1.4-**48**
Linde-Frankl process	林德-弗兰克循环	procédé Linde-Frankl	9.1.4-**49**
Linde high-pressure cycle	林德高压循环	cycle de Linde haute pression	9.1.4-**50**
liquefied natural gas (LNG) *LNG*	液化天然气 (LNG) 液化天然气	gaz naturel liquéfié *GNL*	9.1.4-**51**
liquefied petroleum gas (LPG) *LPG*	液化石油气 (LPG) 液化石油气	gaz de pétrole liquéfié (GPL) *GPL*	9.1.4-**52**
liquid nitrogen (LIN) *LIN*	液氮 (LIN) 液氮	azote liquide	9.1.4-**53**
liquid nitrogen wash	液氮洗涤	lavage à l'azote liquide	9.1.4-**54**
liquid oxygen (LOX) *LOX*	液氧 (LOX) 液氧	oxygène liquide	9.1.4-**55**
liquid-pump plant *internal-compression plant* ◐	液氧泵式装置 内压缩式装置	unité de séparation d'air à pompe *unité de séparation d'air à compression interne* ◐	9.1.4-**56**
LNG *liquefied natural gas (LNG)*	液化天然气 (LNG) 液化天然气	gaz naturel liquéfié *GNL*	9.1.4-**57**
LNG ship of a membrane type	薄膜舱 LNG 船	bateau GNL du type membrane *méthanier du type membrane* ◐	9.1.4-**58**
LNG ship of a Moss type *Moss-Rosenberg-type ship* ◐	Moss 球形舱 LNG 船	bateau GNL du type Moss *méthanier du type Moss* ◐	9.1.4-**59**
LNG ship of a prismatic type	棱形舱 LNG 船	bateau GNL du type prismatique *méthanier à cuves prismatiques* ◐	9.1.4-**60**
loading arm	加载臂	bras de chargement	9.1.4-**61**
low-pressure air-separation plant *low-pressure plant* ◐	低压空气分离装置	unité de séparation d'air basse pression	9.1.4-**62**
low-pressure column *upper column* ◐	低压塔 上塔	colonne basse pression	9.1.4-**63**
low-pressure cycle *low-pressure liquefier* *Kellog cycle* ◐	低压循环	cycle basse pression	9.1.4-**64**

	ENGLISH	汉语	FRANÇAIS
9.1.4-**65**	low-pressure liquefier *low-pressure cycle* *Kellog cycle* ◐	低压循环	cycle basse pression
9.1.4-**66**	low-pressure plant ◐ *low-pressure air-separation plant*	低压空气分离装置	unité de séparation d'air basse pression
9.1.4-**67**	lower column ◐ *high-pressure column* *HP column* ◐	高压塔 下塔	colonne haute pression
9.1.4-**68**	LOX *liquid oxygen (LOX)*	液氧 (LOX) 液氧	oxygène liquide
9.1.4-**69**	LPG *liquefied petroleum gas (LPG)*	液化石油气 (LPG) 液化石油气	gaz de pétrole liquéfié (GPL) *GPL*
9.1.4-**70**	main heat exchanger	主换热器	échangeur de chaleur principal
9.1.4-**71**	medium-pressure air-separation plant *medium-pressure plant* ◐	中压空气分离装置	unité de séparation d'air moyenne pression
9.1.4-**72**	medium-pressure cycle *Claude cycle* ◐	中压循环 克劳德循环中压装置	cycle moyenne pression
9.1.4-**73**	medium-pressure plant ◐ *medium-pressure air-separation plant*	中压空气分离装置	unité de séparation d'air moyenne pression
9.1.4-**74**	methane wash process	甲烷洗涤法	procédé de lavage au méthane
9.1.4-**75**	mixed-component refrigerant ◐ *mixed refrigerant* *multi-component refrigerant* ◐	混合工质	fluide frigorigène mixte
9.1.4-**76**	mixed refrigerant *mixed-component refrigerant* ◐ *multi-component refrigerant* ◐	混合工质	fluide frigorigène mixte
9.1.4-**77**	mixed-refrigerant cycle *inner cascade cycle* *mixed-refrigerant liquefier*	内复叠式循环 内复叠式循环液化器	cycle à fluide frigorigène mixte
9.1.4-**78**	mixed-refrigerant liquefier *mixed-refrigerant cycle* *inner cascade cycle*	内复叠式循环 内复叠式循环液化器	cycle à fluide frigorigène mixte
9.1.4-**79**	MLI *multilayer insulation (MLI)* *superinsulation*	超绝热	superisolation
9.1.4-**80**	molecular sieve	分子筛	tamis moléculaire
9.1.4-**81**	Moss-Rosenberg-type ship ◐ *LNG ship of a Moss type*	Moss 球形舱 LNG 船	bateau GNL du type Moss *méthanier du type Moss* ◐
9.1.4-**82**	multi-component refrigerant ◐ *mixed refrigerant* *mixed-component refrigerant* ◐	混合工质	fluide frigorigène mixte
9.1.4-**83**	multilayer insulation (MLI) *superinsulation* *MLI*	超绝热	superisolation
9.1.4-**84**	natural gas (NG) *NG*	天然气 (NG) 天然气	gaz naturel
9.1.4-**85**	natural gas liquid (NGL) *NGL*	液体天然气 (NGL) 液体天然气	gaz naturel liquide
9.1.4-**86**	NER *boil-off rate (BOR)* *BOR* *evaporation rate* *net evaporation rate (NER)*	蒸发率 (BOR) 蒸发率	taux d'évaporation *taux de pertes par évaporation*

ENGLISH	汉语	FRANÇAIS	
net evaporation rate (NER) *boil-off rate (BOR)* *BOR* *evaporation rate* *NER*	蒸发率 (BOR) 蒸发率	taux d'évaporation *taux de pertes par évaporation*	9.1.4-**87**
NG *natural gas (NG)*	天然气 (NG) 天然气	gaz naturel	9.1.4-**88**
NGL *natural gas liquid (NGL)*	液体天然气 (NGL) 液体天然气	gaz naturel liquide	9.1.4-**89**
nitrogen generator	制氮机	générateur d'azote	9.1.4-**90**
open-rack vaporizer	开架式气化器	regazéifieur à ruissellement d'eau *vaporiseur à ruissellement d'eau*	9.1.4-**91**
oxygen generator	制氧机	générateur d'oxygène	9.1.4-**92**
para- or orthohydrogen	仲氢/正氢	para- ou orthohydrogène	9.1.4-**93**
peak-lopping plant ◐ *peak-shave (LNG) plant* *peak-shaving plant*	LNG 削峰站装置	station d'écrètement de pointe	9.1.4-**94**
peak-shave (LNG) plant *peak-shaving plant* *peak-lopping plant* ◐	LNG 削峰站装置	station d'écrètement de pointe	9.1.4-**95**
peak-shaving plant *peak-shave (LNG) plant* *peak-lopping plant* ◐	LNG 削峰站装置	station d'écrètement de pointe	9.1.4-**96**
plate-and-fin heat exchanger ◐ *brazed-aluminium exchanger*	钎焊铝换热器 板翅式换热器	échangeur de chaleur en aluminium brasé	9.1.4-**97**
poor liquid *shelf liquid* ○	贫氧液空	liquide pauvre	9.1.4-**98**
rare gases	稀有气体	gaz rares	9.1.4-**99**
recondenser	再冷凝器	recondenseur *reliquéfacteur* *réincorporateur* ◐	9.1.4-**100**
reversing heat exchanger	切换式换热器	échangeur de chaleur réversible	9.1.4-**101**
rich liquid	富氧液空	liquide riche	9.1.4-**102**
roll-over effect	涡旋效应	effet de débordement	9.1.4-**103**
satellite station *vaporization station* *customer station*	气化装置	poste de vaporisation	9.1.4-**104**
send-out system *send-out unit*	送出系统 送出机组	unité d'émission	9.1.4-**105**
send-out unit *send-out system*	送出系统 送出机组	unité d'émission	9.1.4-**106**
shelf liquid ○ *poor liquid*	贫氧液空	liquide pauvre	9.1.4-**107**
side-arm column ◐ *argon column* *argon side column* ◐	氩塔	colonne d'argon	9.1.4-**108**
sieve tray	筛板	plateau perforé	9.1.4-**109**
single-column plant *Linde column* ◐	单塔装置	unité à une seule colonne	9.1.4-**110**
sploshing	飞溅	ballotement	9.1.4-**111**
stratification	分层	stratification	9.1.4-**112**

	ENGLISH	汉语	FRANÇAIS
9.1.4-**113**	structured packing	规整填料	garnissage structuré
9.1.4-**114**	submerged combustion vaporizer	浸没式燃烧气化器	regazéifieur à combustion submergée *vaporisateur à combustion submergée*
9.1.4-**115**	sump	液槽	cuve
9.1.4-**116**	superinsulation *multilayer insulation (MLI)* *MLI*	超绝热	superisolation
9.1.4-**117**	trumpler pass *balance stream* ◐	平衡流	flux d'équilibrage
9.1.4-**118**	UHP GAN *ultra-high purity nitrogen (UHP GAN)*	超高纯氮	azote ultrapur
9.1.4-**119**	ultra-high purity nitrogen (UHP GAN) *UHP GAN*	超高纯氮	azote ultrapur
9.1.4-**120**	ultra-high purity oxygen	超高纯氧	oxygène ultrapur
9.1.4-**121**	upper column ◐ *low-pressure column*	低压塔 上塔	colonne basse pression
9.1.4-**122**	vacuum-insulated pipeline	真空绝热管道	gazoduc isolé sous vide
9.1.4-**123**	vacuum-insulated tank cryogenic storage tank	真空绝热罐 低温贮罐	réservoir isolé sous vide
9.1.4-**124**	vaporization station *customer station* *satellite station*	气化装置	poste de vaporisation

	SECTION 9.2 *Cryobiology* **SECTION 9.2.1** *Cryobiology and the influence of low temperatures on living organisms*	节 **9.2** 低温生物学 节 **9.2.1** 低温生物学与低温对活组织（机体）的影响	**SOUS-CHAPITRE 9.2** *Cryobiologie* **SOUS-CHAPITRE 9.2.1** *Cryobiologie et influence des basses températures sur les organismes vivants*
9.2.1-**1**	antifreeze protein	抗冻蛋白	proteine antigel
9.2.1-**2**	biological material	生物材料	matériel biologique
9.2.1-**3**	biological tissue	器官组织	tissu biologique
9.2.1-**4**	bone marrow cell	骨髓细胞	cellule de moelle osseuse
9.2.1-**5**	cellular rupture	细胞破裂 细胞膜破裂	éclatement des cellules
9.2.1-**6**	cellular structure	细胞结构	structure cellulaire
9.2.1-**7**	cold acclimatization	冷适应性	acclimatation au froid
9.2.1-**8**	cold hardening	冷硬化	résistance au froid
9.2.1-**9**	cold preservation	冷藏	conservation à froid
9.2.1-**10**	cold-shock tolerance	冷冲击耐受度	tolérance au choc thermique à froid
9.2.1-**11**	cryo-etching	低温切片	cryodécapage
9.2.1-**12**	cryobiology	低温生物学	cryobiologie
9.2.1-**13**	cryobranding	低温烙印 低温标记	cryomarquage (des animaux)

ENGLISH	汉语	FRANÇAIS	
cryoconservation *cryopreservation*	低温保存	cryoconservation	9.2.1-**14**
cryopreservation *cryoconservation*	低温保存	cryoconservation	9.2.1-**15**
cryoprotectant *cryoprotective agent* *cryoprotector* ◐	低温保护剂	cryoprotecteur *antigel biocompatible* ◐	9.2.1-**16**
cryoprotective agent *cryoprotectant* *cryoprotector* ◐	低温保护剂	cryoprotecteur *antigel biocompatible* ◐	9.2.1-**17**
cryoprotector ◐ *cryoprotectant* *cryoprotective agent*	低温保护剂	cryoprotecteur *antigel biocompatible* ◐	9.2.1-**18**
crystal growth	晶体生长	croissance des cristaux	9.2.1-**19**
crystallization	结晶	cristallisation	9.2.1-**20**
desiccation	除湿 干燥	dessiccation	9.2.1-**21**
devitrification	脱玻作用	dévitrification	9.2.1-**22**
differential scanning calorimetry	差示扫描量热法	calorimétrie différentielle à balayage	9.2.1-**23**
dimethyl sulphoxide	二甲亚砜	diméthyl sulfoxyde	9.2.1-**24**
dormancy	休眠	dormance	9.2.1-**25**
freeze tolerance	冻结耐受力	tolérance à la congélation	9.2.1-**26**
frozen storage	冻结冷藏	conservation à l'état congelé	9.2.1-**27**
glycerol	甘油	glycérol	9.2.1-**28**
heat-shock tolerance	热冲击耐受力	tolérance au choc thermique	9.2.1-**29**
hibernation	冬眠	hibernation	9.2.1-**30**
hypothermal storage	低温冷藏	conservation en état hypothermique	9.2.1-**31**
ice nucleation	冰核晶过程	nucléation de la glace	9.2.1-**32**
intracellular rehydration *intracellular resorption* ◐	细胞内吸收	réhydratation intracellulaire *résorption intracellulaire* ◐	9.2.1-**33**
intracellular resorption ◐ *intracellular rehydration*	细胞内吸收	réhydratation intracellulaire *résorption intracellulaire* ◐	9.2.1-**34**
lag period	滞后期	période de latence	9.2.1-**35**
liposome	脂质体	liposome	9.2.1-**36**
low-temperature hazard	低温危害	dangers des basses températures *risques dus aux basses températures* ◐	9.2.1-**37**
low-temperature survival	低温生存	survie aux basses températures	9.2.1-**38**
microbial flora *microflora* ◐	微生物群	flore microbienne *microflore* ◐	9.2.1-**39**
microflora ◐ *microbial flora*	微生物群	flore microbienne *microflore* ◐	9.2.1-**40**
osmolarity	重量克分子渗透压浓度 （同渗重摩，渗透度，溶质度）	osmolarité	9.2.1-**41**
osmosis	渗透	osmose	9.2.1-**42**
osmotic shock	渗透冲击	choc osmotique	9.2.1-**43**
recrystallization	再结晶	recristallisation	9.2.1-**44**

	ENGLISH	汉语	FRANÇAIS
9.2.1-**45**	rehydration	再水合作用	réhydratation
9.2.1-**46**	rewarming	再升温	réchauffement
9.2.1-**47**	vernalization	春化作用	printanisation *vernalisation* ◐
9.2.1-**48**	viability	生命力	viabilité
9.2.1-**49**	vitrification	玻化	vitrification
9.2.1-**50**	warming	升温	réchauffement
9.2.1-**51**	warming rate	升温速度	vitesse de réchauffement

	SECTION 9.2.2 *Freeze-drying*	节 **9.2.2** 冻干	**SOUS-CHAPITRE 9.2.2** *Lyophilisation*
9.2.2-**1**	atmospheric freeze-drying	常压冷冻干燥	cryodessiccation atmosphérique
9.2.2-**2**	batch freeze-drying *discontinuous freeze-drying*	间歇冷冻干燥	lyophilisation discontinue *lyophilisation charge par charge* ◐
9.2.2-**3**	centrifugal freeze-drying	离心冷冻干燥	lyophilisation avec centrifugation
9.2.2-**4**	continuous freeze-drying	连续冷冻干燥	lyophilisation continue
9.2.2-**5**	desiccation ratio	脱水率	taux de dessiccation
9.2.2-**6**	desorbable water ◐ *releasable water*	可释水	eau libérable
9.2.2-**7**	diffuse sublimation front ◐	扩散升华界面	front de sublimation diffus ◐
9.2.2-**8**	discontinuous freeze-drying *batch freeze-drying*	间歇冷冻干燥	lyophilisation discontinue *lyophilisation charge par charge* ◐
9.2.2-**9**	drum freeze-drier	筒状冷冻干燥器	lyophilisateur à tambour
9.2.2-**10**	dryness ratio	干物质比	taux de siccité
9.2.2-**11**	excipient *freeze-drying additive*	冷冻干燥添加剂	support de lyophilisation *additif pour lyophilisation*
9.2.2-**12**	extraction ratio	提取率	taux d'extraction
9.2.2-**13**	final temperature of freezing	最终冻结温度	température de congélation totale
9.2.2-**14**	freeze-dried *lyophilized*	冷冻干燥 升华干燥处理的 冻干的	lyophilisé
9.2.2-**15**	freeze-drier	冷冻干燥机	lyophilisateur
9.2.2-**16**	freeze-drying *lyophilization*	冷冻干燥法 升华干燥法	lyophilisation *cryodessiccation* ◐
9.2.2-**17**	freeze-drying additive *excipient*	冷冻干燥添加剂	support de lyophilisation *additif pour lyophilisation*
9.2.2-**18**	lyophilic	亲液的	lyophile
9.2.2-**19**	lyophilizate	升华干燥产品	lyophilisat
9.2.2-**20**	lyophilization *freeze-drying*	冷冻干燥法 升华干燥法	lyophilisation *cryodessiccation* ◐
9.2.2-**21**	lyophilized *freeze-dried*	冷冻干燥 升华干燥处理的 冻干的	lyophilisé

ENGLISH	汉语	FRANÇAIS	
manifold drying apparatus	歧管干燥器	hérisson	9.2.2-**22**
powder freezing	粉状冻结	congélation en poudre	9.2.2-**23**
primary drying	初次干燥	dessiccation primaire	9.2.2-**24**
rate of sublimation	升华率	taux de sublimation	9.2.2-**25**
releasable water *desorbable water* ◐	可释水	eau libérable	9.2.2-**26**
residual moisture	剩余湿分	humidité résiduelle	9.2.2-**27**
residual pressure	剩余压力	pression résiduelle	9.2.2-**28**
secondary drying	二次干燥	dessiccation secondaire	9.2.2-**29**
sharp sublimation front	清晰升华界面	front net de sublimation *front aigu de sublimation* ◐	9.2.2-**30**
shell freezing	壳层冻结	congélation en coquille	9.2.2-**31**
spray freeze-drying	喷雾冷冻干燥	lyophilisation par pulvérisation	9.2.2-**32**
sublimation front *sublimation interface*	升华界面	front de sublimation *interface* ◐	9.2.2-**33**
sublimation interface *sublimation front*	升华界面	front de sublimation *interface* ◐	9.2.2-**34**
sublimer	升华干燥箱 冻干室	sublimateur	9.2.2-**35**
tray drying chamber	盘式干燥箱	chambre à vide à étagères *chambre de dessiccation à étagères* ◐	9.2.2-**36**
vacuum freezing	真空冻结	congélation par le vide	9.2.2-**37**
vibrating freeze-drier	振动冷冻干燥机	lyophilisateur à vibreur	9.2.2-**38**
water holding capacity	保水能力 持水量	capacité de rétention d'eau	9.2.2-**39**

SECTION 9.2.3 *Cryomedicine*	节 **9.2.3** 低温医学	**SOUS-CHAPITRE 9.2.3** *Cryomédecine*	
(artificial) hibernation *hypothermia*	（人工）冬眠 人体降温	hypothermie *hibernation (artificielle)* ◐	9.2.3-**1**
biomaterial	生物材料	matériel biologique	9.2.3-**2**
blood bank	血库	banque de sang	9.2.3-**3**
bone bank	骨库	banque d'os	9.2.3-**4**
cadaver storage *mortuary*	太平间	morgue	9.2.3-**5**
cooling cannula *cryogenic needle* *cryoprobe* *cryosurgical probe*	低温手术探针	cryosonde	9.2.3-**6**
cryo-ophthalmology	低温眼科学	cryo-ophtalmologie	9.2.3-**7**
cryoablation *cryoextirpation*	低温切除 低温摘除	cryoablation	9.2.3-**8**
cryoadherence *cryoadhesion*	低温粘连	cryoadhérence	9.2.3-**9**

	ENGLISH	汉语	FRANÇAIS
9.2.3-**10**	cryoadhesion *cryoadherence*	低温粘连	cryoadhérence
9.2.3-**11**	cryoanalgesia	低温止痛	cryoanalgésie
9.2.3-**12**	cryoapplication	低温涂敷	cryoapplication
9.2.3-**13**	cryocauterization	低温烧灼	cryocautérisation
9.2.3-**14**	cryocauterizer	低温烧灼器	cryocautère
9.2.3-**15**	cryocoagulation	低温凝结	cryocoagulation
9.2.3-**16**	cryodestruction	低温破坏	cryodestruction
9.2.3-**17**	cryoextirpation *cryoablation*	低温切除 低温摘除	cryoablation
9.2.3-**18**	cryoextraction	低温切除	cryoextraction
9.2.3-**19**	cryofixation *cryopexy* ○	低温固定	cryofixation *cryopexie* ○
9.2.3-**20**	cryogenic needle *cooling cannula* *cryoprobe* *cryosurgical probe*	低温手术探针	cryosonde
9.2.3-**21**	cryoimmunology	冷冻免疫学	cryoimmunologie
9.2.3-**22**	cryolesion	低温损伤	cryolésion
9.2.3-**23**	cryomedicine	低温医学	cryomédecine
9.2.3-**24**	cryometer	低温计/低温温度计	cryothermomètre
9.2.3-**25**	cryopexy ○ *cryofixation*	低温固定	cryofixation *cryopexie* ○
9.2.3-**26**	cryoprobe *cooling cannula* *cryogenic needle* *cryosurgical probe*	低温手术探针	cryosonde
9.2.3-**27**	cryoresection	低温切除	cryorésection
9.2.3-**28**	cryoretinopexy	视网膜低温固定	cryorétinopexie
9.2.3-**29**	cryoscalpel *cryostylet*	低温手术刀	cryobistouri
9.2.3-**30**	cryostylet *cryoscalpel*	低温手术刀	cryobistouri
9.2.3-**31**	cryosurgery	低温手术	cryochirurgie
9.2.3-**32**	cryosurgical probe *cooling cannula* *cryogenic needle* *cryoprobe*	低温手术探针	cryosonde
9.2.3-**33**	cryotherapy *cryotreatment* ◐	低温疗法 低温治疗	cryothérapie *cryotraitement* ◐
9.2.3-**34**	cryotool	低温器械	cryo-outil
9.2.3-**35**	cryotreatment ◐ *cryotherapy*	低温疗法 低温治疗	cryothérapie *cryotraitement* ◐
9.2.3-**36**	graft *transplant*	移植	greffon
9.2.3-**37**	hypothermia *(artificial) hibernation*	(人工)冬眠 人体降温	hypothermie *hibernation (artificielle)* ◐
9.2.3-**38**	hypothermic blanket	低温褥	couverture réfrigérante

ENGLISH	汉语	FRANÇAIS	
mortuary *cadaver storage*	太平间	morgue	9.2.3-**39**
tissue bank	组织库	banque de tissus	9.2.3-**40**
transplant *graft*	移植	greffon	9.2.3-**41**
transplantation	移植	transplantation	9.2.3-**42**
University of Wisconsin solution	威斯康星大学溶液	solution de conservation UW	9.2.3-**43**

章 10. 制冷的其它应用

◐ 容许（或默许）术语

○ 过时术语

ENGLISH	汉语	FRANÇAIS	
SECTION 10.1 *Water ice*	节 **10.1** 水冰	**SOUS-CHAPITRE 10.1** *Glace hydrique*	
block ice *cake ice* ◐	块冰	glace en bloc *glace en pain* *glace en mouleaux* ◐	10.1-**1**
briquette ice ○	砖冰	glace en briquettes ○	10.1-**2**
broken ice *lump ice* ◐	碎冰块	glace concassée	10.1-**3**
cake ice ◐ *block ice*	块冰	glace en bloc *glace en pain* *glace en mouleaux* ◐	10.1-**4**
chip ice *chipped ice* ◐	片冰	glace en copeaux	10.1-**5**
chipped ice ◐ *chip ice*	片冰	glace en copeaux	10.1-**6**
clear ice	透明冰	glace transparente	10.1-**7**
crushed ice	碎冰	glace broyée	10.1-**8**
crystal ice	结晶冰	glace cristal	10.1-**9**
cube ice	方块冰	glace en cubes	10.1-**10**
flake ice *slice ice* *scale ice* ◐	片冰	glace en écailles *glace en éclats* ◐ *glace en flocons* ◐	10.1-**11**
ice (to)	加冰（动词）	glacer	10.1-**12**
ice block *ice cake* ◐	冰块	bloc de glace *pain de glace* ◐ *mouleau de glace* ◐	10.1-**13**
ice cake ◐ *ice block*	冰块	bloc de glace *pain de glace* ◐ *mouleau de glace* ◐	10.1-**14**
ice cube	方块冰	cube de glace *glaçon*	10.1-**15**
lump ice ◐ *broken ice*	碎冰块	glace concassée	10.1-**16**
opaque ice *white ice*	白冰	glace opaque	10.1-**17**
plate ice	板冰	glace en plaques	10.1-**18**
processed ice	加工冰	glace fractionnée	10.1-**19**
ribbon ice	带状冰	glace en ruban	10.1-**20**
scale ice ◐ *flake ice* *slice ice*	片冰	glace en écailles *glace en éclats* ◐ *glace en flocons* ◐	10.1-**21**
sea-water ice	海水冰	glace d'eau de mer	10.1-**22**
shell ice ○ *tube ice*	管 冰	glace en tubes	10.1-**23**
sized ice	规格冰	glace calibrée	10.1-**24**
slice ice *flake ice* *scale ice* ◐	片冰	glace en écailles *glace en éclats* ◐ *glace en flocons* ◐	10.1-**25**
slush	雪水	neige fondante	10.1-**26**

	ENGLISH	汉语	FRANÇAIS
10.1-**27**	slush ice	湿雪冰	glace-neige mouillée
10.1-**28**	small ice	小块冰	glace divisée
10.1-**29**	snow	雪	neige
10.1-**30**	snow-ice	雪冰	glace-neige
10.1-**31**	tube ice *shell ice* ○	管 冰	glace en tubes
10.1-**32**	(water) ice	(水)冰	glace (hydrique)
10.1-**33**	white ice *opaque ice*	白冰	glace opaque

	SECTION 10.2 *Ice-making plants*	节 **10.2** 制冰装置	**SOUS-CHAPITRE 10.2** *Fabriques de glace*
10.2-**1**	air agitation (in ice making)	空气搅拌（制冰中）	insufflation d'air (dans les mouleaux à glace)
10.2-**2**	brine tank ◐ *ice-making tank* *ice tank* ◐	制冰池	bac à glace *bac à saumure* *bac générateur de glace*
10.2-**3**	can dump (USA) ◐ *ice tip*	倒冰架	basculeur de mouleaux à glace *culbuteur de mouleaux à glace*
10.2-**4**	can filler	加水器	emplisseur (de mouleaux) *herse de remplissage* ◐ *remplisseur (de mouleaux)* ◐
10.2-**5**	core pulling and filling system	换水器	dispositif de succion du noyau
10.2-**6**	dip tank ◐ *thawing tank*	融冰槽	bac de démoulage
10.2-**7**	freezing cylinder ◐ *ice generator* *freezing drum* ◐	成冰器	organe générateur de glace
10.2-**8**	freezing drum ◐ *ice generator* *freezing cylinder* ◐	成冰器	organe générateur de glace
10.2-**9**	ice can ◐ *ice mould*	冰桶	mouleau (à glace)
10.2-**10**	ice can frame ◐ *ice mould frame* *ice can grid* ◐	冰桶架	châssis pour mouleaux
10.2-**11**	ice can grid ◐ *ice mould frame* *ice can frame* ◐	冰桶架	châssis pour mouleaux
10.2-**12**	ice can group ◐ *row of moulds* *row of cans* ◐	冰模排	rangée de mouleaux
10.2-**13**	ice can truck	吊冰桶车	treuil mobile pour mouleaux
10.2-**14**	ice cellar *snow cellar* ◐	冰窖	glacière
10.2-**15**	ice chute	滑冰道	glissière à glace
10.2-**16**	ice crane	吊冰行车	pont roulant de bac à glace
10.2-**17**	ice crusher	碎冰机	broyeur à glace

ENGLISH	汉语	FRANÇAIS	
ice dump table	倒冰台	table de démoulage	10.2-**18**
ice factory ◐ *ice-making plant*	制冰厂	fabrique de glace	10.2-**19**
ice generator *freezing cylinder* ◐ *freezing drum* ◐	成冰器	organe générateur de glace	10.2-**20**
ice-maker	制冰机	générateur de glace	10.2-**21**
ice-making plant *ice factory* ◐	制冰厂	fabrique de glace	10.2-**22**
ice-making tank *brine tank* ◐ *ice tank* ◐	制冰池	bac à glace *bac à saumure* *bac générateur de glace*	10.2-**23**
ice manufacture	制冰	fabrication de la glace	10.2-**24**
ice mould *ice can* ◐	冰桶	mouleau (à glace)	10.2-**25**
ice mould frame *ice can frame* ◐ *ice can grid* ◐	冰桶架	châssis pour mouleaux	10.2-**26**
ice storage room	冰库	réserve à glace *resserre à glace* *glacière* ◐	10.2-**27**
ice tank ◐ *ice-making tank* *brine tank* ◐	制冰池	bac à glace *bac à saumure* *bac générateur de glace*	10.2-**28**
ice tip *can dump (USA)* ◐	倒冰架	basculeur de mouleaux à glace *culbuteur de mouleaux à glace*	10.2-**29**
row of cans ◐ *row of moulds* *ice can group* ◐	冰模排	rangée de mouleaux	10.2-**30**
row of moulds *ice can group* ◐ *row of cans* ◐	冰模排	rangée de mouleaux	10.2-**31**
snow cellar ◐ *ice cellar*	冰窖	glacière	10.2-**32**
thawing tank *dip tank* ◐	融冰槽	bac de démoulage	10.2-**33**
water forecooler ◐ *water precooler*	水预冷器	prérefroidisseur d'eau	10.2-**34**
water precooler *water forecooler* ◐	水预冷器	prérefroidisseur d'eau	10.2-**35**

SECTION 10.3 *Winter sports*	节 **10.3** 冬季运动	**SOUS-CHAPITRE 10.3** *Sports d'hiver*	
curling rink	冰壶冰场	piste de curling	10.3-**1**
de-icing	融冰	fusion du plateau de glace	10.3-**2**
ice hockey rink	冰球场	patinoire pour hockey	10.3-**3**
ice rink ◐ *skating rink*	滑冰场	patinoire	10.3-**4**
ice slab	冰板	plateau de glace	10.3-**5**

	ENGLISH	汉语	FRANÇAIS
10.3-6	olympic rink	奥林匹克滑冰场	patinoire olympique
10.3-7	pipe (cooling) grids	（冰场）制冷排管	réseau de tubes (refroidisseurs)
10.3-8	rink floor	滑冰场底板	soubassement de patinoire
10.3-9	skating rink *ice rink* ◐	滑冰场	patinoire
10.3-10	snow gun *snow maker*	雪枪	canon à neige *enneigeur* ◐
10.3-11	snow maker *snow gun*	雪枪	canon à neige *enneigeur* ◐
10.3-12	speed track	速滑道	anneau de vitesse
10.3-13	surfacing (of ice)	冰面处理	surfaçage (de la glace)

	SECTION 10.4 *Chemical industries*	节 **10.4** 化学工业	**SOUS-CHAPITRE 10.4** *Industries chimiques*
10.4-1	brackish water	淡盐水	eau saumâtre
10.4-2	cryoconcentration *freeze-concentration* ◐	冻结浓缩	cryoconcentration *concentration par congélation* ◐
10.4-3	cryogrinding *freeze-grinding* ◐	低温研磨	cryobroyage
10.4-4	cryotrimming *cryotumbling* ◐	低温整理	cryoébarbage
10.4-5	cryotumbling ◐ *cryotrimming*	低温整理	cryoébarbage
10.4-6	freeze-concentration ◐ *cryoconcentration*	冻结浓缩	cryoconcentration *concentration par congélation* ◐
10.4-7	freeze-grinding ◐ *cryogrinding*	低温研磨	cryobroyage
10.4-8	freeze desalination	冻结脱盐	dessalement par congélation
10.4-9	freeze out (to)	冻析（动词）	séparer par congélation
10.4-10	soft water	软水	eau douce

	SECTION 10.5 *Metallurgy and mechanical industries*	节 **10.5** 冶金工业与机械工业	**SOUS-CHAPITRE 10.5** *Industries métallurgiques et mécaniques*
10.5-1	ageing (of materials)	(材料的)时效	vieillissement (des matériaux)
10.5-2	cold brittleness	冷脆性	fragilité au froid
10.5-3	cold-dimensional stabilization	冷尺寸稳定	stabilisation dimensionnelle par le froid
10.5-4	cold-shrink fitting *expansion fitting* ◐	冷缩装配	assemblage par contraction *emmanchement à froid*
10.5-5	cryohardening	低温硬化	cryotrempe
10.5-6	expansion fitting ◐ *cold shrink fitting*	冷缩装配	assemblage par contraction *emmanchement à froid*

ENGLISH	汉语	FRANÇAIS	
hardening *quenching* ◐	淬火	trempe	10.5-**7**
quenching ◐ *hardening*	淬火	trempe	10.5-**8**
shrink disassembly	收缩拆卸	démontage par contraction	10.5-**9**

SECTION 10.6 *Miscellaneous applications of refrigeration*	节 **10.6** 制冷的其余应用	**SOUS-CHAPITRE 10.6** *Applications diverses du froid*	
aggregate cooling	混凝土骨料冷却	refroidissement des agrégats	10.6-**1**
concrete-dam cooling	混凝土坝冷却	refroidissement des barrages en béton	10.6-**2**
embedded cooling coils	埋置式冷却盘管	tubes de refroidissement noyés	10.6-**3**
freezing the soil	冻结土壤	congélation du sol *congélation du terrain* ◐	10.6-**4**
frozen earth storage ◐ *frozen ground storage* *frozen soil storage* ◐	冻土地下贮藏	stockage en terrain congelé *stockage en terre gelée* ◐	10.6-**5**
frozen ground storage *frozen earth storage* ◐ *frozen soil storage* ◐	冻土地下贮藏	stockage en terrain congelé *stockage en terre gelée* ◐	10.6-**6**
frozen soil storage ◐ *frozen ground storage* *frozen earth storage* ◐	冻土地下贮藏	stockage en terrain congelé *stockage en terre gelée* ◐	10.6-**7**
permafrost	永久冻土	pergélisol *permagel* ◐	10.6-**8**
simulation chamber	模拟室	chambre de simulation	10.6-**9**
space simulator	空间模拟器	simulateur spatial	10.6-**10**
thermal shroud	热窖	enveloppe de protection thermique	10.6-**11**

章 11. 制冷与环境

◐ 容许（或默许）术语

○ 过时术语

ENGLISH	汉语	FRANÇAIS	
CHAPTER 11 *Refrigeration and the environment*	章 **11** 制冷与环境	**CHAPITRE 11** *Froid et environnement*	
Article 5 country	第五条款国家	pays de l'Article 5	**11-1**
atmospheric lifetime	在大气中持续时间	durée de vie atmosphérique	**11-2**
climate change	气候变化	changement climatique	**11-3**
climate change attributed to human activity	人类活动引起的气候变化	changement climatique attribué à l'activité humaine	**11-4**
Conference of the Parties *COP*	成员国大会	Conférence des Parties	**11-5**
controlled substance	受控物质	substance réglementée *substance sous contrôle* ◐	**11-6**
COP *Conference of the Parties*	成员国大会	Conférence des Parties	**11-7**
direct warming impact	（气候）变暖直接影响	impact direct sur le réchauffement planétaire *impact climatique*	**11-8**
drop-in replacement	直接替代	remplacement immédiat *drop-in* ◐	**11-9**
fluorocarbon	氟碳化合物（氟代烃，含 CFCs, HCFCs, HFCs, FC）	fluorocarbure	**11-10**
global environmental issues	全球环境问题	enjeux climatiques planétaires	**11-11**
global warming *greenhouse effect*	全球（气候）变暖 温室效应	réchauffement planétaire *effet de serre*	**11-12**
global warming potential *GWP*	全球变暖潜值	potentiel de réchauffement planétaire *GWP*	**11-13**
greenhouse effect *global warming*	全球（气候）变暖 温室效应	réchauffement planétaire *effet de serre*	**11-14**
greenhouse gas	温室气体	gaz à effet de serre	**11-15**
GWP *global warming potential*	全球变暖潜值	potentiel de réchauffement planétaire *GWP*	**11-16**
halocarbon	卤代烃	hydrocarbure halogéné *halocarbure*	**11-17**
halon	哈龙	halon	**11-18**
indirect warming impact	（气候）变暖间接影响	impact indirect sur le réchauffement planétaire	**11-19**
Kyoto Protocol	京都议定书	Protocole de Kyoto	**11-20**
life-cycle cost analysis	生命周期费用分析	analyse du coût (d'une installation) pendant son cycle de vie	**11-21**
Life-Cycle Climate Performance (LCCP)	寿命期气候性能	impact sur le climat au cours du cycle de vie	**11-22**
Montreal Protocol	蒙特利尔议定书	Protocole de Montréal	**11-23**
ODP *ozone depletion potential*	臭氧损耗潜值	potentiel d'appauvrissement de la couche d'ozone *ODP*	**11-24**
ODS *ozone-depleting substance*	损耗臭氧的物质	substance appauvrissant l'ozone	**11-25**
ozone	臭氧	ozone	**11-26**

	ENGLISH	汉语	FRANÇAIS
11-27	ozone-depleting substance *ODS*	损耗臭氧的物质	substance appauvrissant l'ozone
11-28	ozone depletion	臭氧损耗	appauvrissement de l'ozone
11-29	ozone depletion potential *ODP*	臭氧损耗潜值	potentiel d'appauvrissement de la couche d'ozone *ODP*
11-30	ozone layer	臭氧层	couche d'ozone
11-31	phase-out	淘汰	élimination
11-32	radiative forcing	（对气候的）辐射	forçage radiatif
11-33	retrofit	改型 更新	conversion
11-34	stratosphere	同温（平流）层	stratosphère
11-35	TEWI *Total Equivalent Warming Impact*	变暖影响总当量	TEWI
11-36	Total Equivalent Warming Impact *TEWI*	变暖影响总当量	TEWI
11-37	transitional substance	过渡物质	frigorigène de transition
11-38	ultraviolet radiation	紫外线辐射	rayonnement ultraviolet

按字母顺序编排的英文索引 ALPHABETICAL INDEX: ENGLISH

◐ 容许（或默许）术语 tolerated term
○ 过时术语 outdated term

T

U

V

W

Y

Z